STATISTICAL SOFTWARE ENGINEERING

Panel on Statistical Methods in Software Engineering
Committee on Applied and Theoretical Statistics
Board on Mathematical Sciences
Commission on Physical Sciences, Mathematics, and Applications
National Research Council

National Academy Press
Washington, D.C. 1996

NOTICE: The project that is the subject of this report was approved by the Governing Board of the National Research Council, whose members are drawn from the councils of the National Academy of Sciences, the National Academy of Engineering, and the Institute of Medicine.

The National Academy of Sciences is a private, nonprofit, self-perpetuating society of distinguished scholars engaged in scientific and engineering research, dedicated to the furtherance of science and technology and to their use for the general welfare. Upon the authority of the charter granted to it by the Congress in 1863, the Academy has a mandate that requires it to advise the federal government on scientific and technical matters. Dr. Bruce Alberts is president of the National Academy of Sciences.

The National Academy of Engineering was established in 1964, under the charter of the National Academy of Sciences, as a parallel organization of outstanding engineers. It is autonomous in its administration and in the selection of its members, sharing with the National Academy of Sciences the responsibility for advising the federal government. The National Academy of Engineering also sponsors engineering programs aimed at meeting national needs, encourages education and research, and recognizes the superior achievement of engineers. Dr. Harold Liebowitz is president of the National Academy of Engineering.

The Institute of Medicine was established in 1970 by the National Academy of Sciences to secure the services of eminent members of appropriate professions in the examination of policy matters pertaining to the health of the public. The Institute acts under the responsibility given to the National Academy of Sciences by its congressional charter to be an adviser to the federal government and, upon its own initiative, to identify issues of medical care, research, and education. Dr. Kenneth I. Shine is president of the Institute of Medicine.

The National Research Council was organized by the National Academy of Sciences in 1916 to associate the broad community of science and technology with the Academy's purposes of furthering knowledge and advising the federal government. Functioning in accordance with general policies determined by the Academy, the Council has become the principal operating agency of both the National Academy of Sciences and the National Academy of Engineering in providing services to the government, the public, and the scientific and engineering communities. The Council is administered jointly by both Academies and the Institute of Medicine. Dr. Bruce Alberts and Dr. Harold Liebowitz are chairman and vice-chairman, respectively, of the National Research Council.

This project was supported by the Advanced Research Projects Agency, Army Research Office, National Science Foundation, and Department of the Navy's Office of the Chief of Naval Research. Any opinions, findings, and conclusions or recommendations expressed in this material are those of the authors and do not necessarily reflect the views of the sponsors. Furthermore, the content of the report does not necessarily reflect the position or the policy of the U.S. government, and no official endorsement should be inferred.

Copyright 1996 by the National Academy of Sciences. All rights reserved.

Library of Congress Catalog Card Number 95-71101
International Standard Book Number 0-309-05344-7

Additional copies of this report are available from: National Academy Press, Box 285
2101 Constitution Avenue, N.W.
Washington, D.C. 20055
800-624-6242
202-334-3313 (in the Washington metropolitan area)
B-676

Printed in the United States of America

PANEL ON STATISTICAL METHODS IN SOFTWARE ENGINEERING

DARYL PREGIBON, AT&T Bell Laboratories, *Chair*
HERMAN CHERNOFF, Harvard University
BILL CURTIS, Carnegie Mellon University
SIDDHARTHA R. DALAL, Bellcore
GLORIA J. DAVIS, NASA-Ames Research Center
RICHARD A. DEMILLO, Bellcore
STEPHEN G. EICK, AT&T Bell Laboratories
BEV LITTLEWOOD, City University, London, England
CHITOOR V. RAMAMOORTHY, University of California, Berkeley

Staff
JOHN R. TUCKER, Director

COMMITTEE ON APPLIED AND THEORETICAL STATISTICS

JON R. KETTENRING, Bellcore, *Chair*
RICHARD A. BERK, University of California, Los Angeles
LAWRENCE D. BROWN, University of Pennsylvania
NICHOLAS P. JEWELL, University of California, Berkeley
JAMES D. KUELBS, University of Wisconsin
JOHN LEHOCZKY, Carnegie Mellon University
DARYL PREGIBON, AT&T Bell Laboratories
FRITZ SCHEUREN, George Washington University
J. LAURIE SNELL, Dartmouth College
ELIZABETH THOMPSON, University of Washington

Staff
JACK ALEXANDER, Program Officer

BOARD ON MATHEMATICAL SCIENCES

AVNER FRIEDMAN, University of Minnesota, *Chair*
LOUIS AUSLANDER, City University of New York
HYMAN BASS, Columbia University
MARY ELLEN BOCK, Purdue University
PETER E. CASTRO, Eastman Kodak Company
FAN R.K. CHUNG, University of Pennsylvania
R. DUNCAN LUCE, University of California, Irvine
SUSAN MONTGOMERY, University of Southern California
GEORGE NEMHAUSER, Georgia Institute of Technology
ANIL NERODE, Cornell University
INGRAM OLKIN, Stanford University
RONALD F. PEIERLS, Brookhaven National Laboratory
DONALD ST. P. RICHARDS, University of Virginia
MARY F. WHEELER, Rice University
WILLIAM P. ZIEMER, Indiana University

Ex Officio Member
JON R. KETTENRING, Bellcore
　Chair, Committee on Applied and Theoretical Statistics

Staff
JOHN R. TUCKER, Director
JACK ALEXANDER, Program Officer
RUTH E. O'BRIEN, Staff Associate
BARBARA W. WRIGHT, Administrative Assistant

COMMISSION ON PHYSICAL SCIENCES, MATHEMATICS, AND APPLICATIONS

ROBERT J. HERMANN, United Technologies Corporation, *Chair*
STEPHEN L. ADLER, Institute for Advanced Study
PETER M. BANKS, Environmental Research Institute of Michigan
SYLVIA T. CEYER, Massachusetts Institute of Technology
L. LOUIS HEGEDUS, W.R. Grace and Company
JOHN E. HOPCROFT, Cornell University
RHONDA J. HUGHES, Bryn Mawr College
SHIRLEY A. JACKSON, U.S. Nuclear Regulatory Commission
KENNETH I. KELLERMANN, National Radio Astronomy Observatory
KEN KENNEDY, Rice University
THOMAS A. PRINCE, California Institute of Technology
JEROME SACKS, National Institute of Statistical Sciences
L.E. SCRIVEN, University of Minnesota
LEON T. SILVER, California Institute of Technology
CHARLES P. SLICHTER, University of Illinois at Urbana-Champaign
ALVIN W. TRIVELPIECE, Oak Ridge National Laboratory
SHMUEL WINOGRAD, IBM T.J. Watson Research Center
CHARLES A. ZRAKET, Mitre Corporation (retired)

NORMAN METZGER, Executive Director

Preface

The development and the production of high-quality, reliable, complex computer software have become critical issues in the enormous worldwide computer technology market. The capability to efficiently engineer computer software development and production processes is central to the future economic strength, competitiveness, and national security of the United States. However, problems related to software quality, reliability, and safety persist, a prominent example being the failure on several occasions of major local and national telecommunications networks. It is now acknowledged that the costs of producing and maintaining software greatly exceed the costs of developing, producing, and maintaining hardware. Thus the development and application of cost-saving tools, along with techniques for ensuring quality and reliability in software engineering, are primary goals in today's software industry. The enormity of this software production and maintenance activity is such that any tools contributing to serious cost savings will yield a tremendous payoff in absolute terms.

At a meeting of the Committee on Applied and Theoretical Statistics (CATS) of the National Research Council (NRC), participants identified software engineering as an area presenting numerous opportunities for fruitful contributions from statistics and offering excellent potential for beneficial interactions between statisticians and software engineers that might promote improved software engineering practice and cost savings. To delineate these opportunities and focus attention on contexts promising useful interactions, CATS convened a study panel to gather information and produce a report that would (1) exhibit improved methods for assessing software productivity, quality, reliability, associated risk, and safety and for managing software development processes, (2) outline a program of research in the statistical sciences and their applications to software engineering with the aim of motivating and attracting new researchers from the mathematical sciences, statistics, and software engineering fields to tackle these important and pressing problem areas, and (3) emphasize the relevance of using rigorous statistical and probabilistic techniques in software engineering contexts and suggest opportunities for further research in this direction.

To help identify important issues and obtain a broad range of perspectives on them, the panel organized an information-gathering forum on October 11-12, 1993, at which 12 invited speakers addressed how statistical methods impinge on the software development process, software metrics, software dependability and testing, and software visualization. The forum also included consideration of nonstandard methods and select case studies (see the forum program in the appendix). The panel hopes that its report, which is based on the panel's expertise as well as information presented at the forum, will contribute to positive advances in software engineering and, as a subsidiary benefit, be a stimulus for other closely related disciplines, e.g., applied mathematics, operations research, computer science, and systems and industrial engineering. The panel is, in fact, very enthusiastic about the opportunities facing the statistical community and hopes to convey this enthusiasm in this report.

The panel gratefully acknowledges the assistance and information provided by a number of individuals, including the 12 forum speakers—T.W. Keller, D. Card, V.R. Basili, J.C. Munson, J.C. Knight, R. Lipton, T. Yamaura, S. Zweben, M.S. Phadke, E.E. Sumner, Jr., W. Hill, and J. Stasko—four anonymous reviewers, the NRC staff of the Board on Mathematical Sciences who supported the various facets of this project, and Susan Maurizi for her work in editing the manuscript.

Contents

EXECUTIVE SUMMARY 1

1 INTRODUCTION 5

2 CASE STUDY: NASA SPACE SHUTTLE FLIGHT CONTROL SOFTWARE 9
    Overview of Requirements 9
    The Operational Life Cycle 10
    A Statistical Approach to Managing the Software Production Process 10
        Fault Detection 11
        Safety Certification 12

3 A SOFTWARE PRODUCTION MODEL 13
    Problem Formulation and Specification of Requirements 14
    Design 14
    Implementation 16
    Testing 18

4 CRITIQUE OF SOME CURRENT APPLICATIONS OF STATISTICS IN
   SOFTWARE ENGINEERING 27
    Cost Estimation 27
        Statistical Inadequacies in Estimating 29
        Process Volatility 30
        Maturity and Data Granularity 30
        Reliability of Model Inputs 31
        Managing to Estimates 32
    Assessment and Reliability 32
        Reliability Growth Modeling 32
        Influence of the Development Process on Software Dependability 36
        Influence of the Operational Environment on Software Dependability 37
        Safety-Critical Software and the Problem of Assuring Ultrahigh Dependability 38
            Design Diversity, Fault Tolerance, and General Issues of Dependence 38
            Judgment and Decision-making Framework 39
            Structural Modeling Issues 40
        Experimentation, Data Collection, and General Statistical Techniques 40
    Software Measurement and Metrics 41

5 STATISTICAL CHALLENGES 43
    Software Engineering Experimental Issues 43
    Combining Information 46
    Visualization in Software Engineering 48

    Configuration Management Data 49
    Function Call Graphs 50
    Test Code Coverage 50
    Code Metrics 50
    Challenges for Visualization 52
    Opportunities for Visualization 52
  Orthogonal Defect Classification 59

6 SUMMARY AND CONCLUSIONS 61
  Institutional Model for Research 62
  Model for Data Collection and Analysis 62
  Issues in Education 64

REFERENCES 67

APPENDIX: FORUM PROGRAM 72

Executive Summary

Software, a critical core industry that is essential to U.S. interests in science, technology, and defense, is ubiquitous in today's society. Software coexists with hardware in our transportation, communication, financial, and medical systems. As these systems grow in size and complexity and our dependence on them increases, the need to ensure software reliability and safety, fault tolerance, and dependability becomes paramount. Building software is now viewed as an engineering discipline, *software engineering,* which aims to develop methodologies and procedures to control the whole software development process. Besides the issue of controlling and improving software quality, the issue of improving the productivity of the software development process is also becoming important from the industrial perspective.

PURPOSE AND SCOPE OF THIS STUDY

Although statistical methods have a long history of contributing to improved practices in manufacturing and in traditional areas of science, technology, and medicine, they have up to now had little impact on software development processes. This report attempts to bridge the islands of knowledge and experience between statistics and software engineering by enunciating a new interdisciplinary field: *statistical software engineering*. It is hoped that the report will help seed the field of statistical software engineering by indicating opportunities for statistical thinking to contribute to increased understanding of software and software production, and thereby enhance the quality and productivity of both.

This report is the result of a study by a panel convened by the Committee on Applied and Theoretical Statistics (CATS), a standing committee of the Board on Mathematical Sciences of the National Research Council, to identify challenges and opportunities in the development and implementation of software involving significant statistical content. In addition to pointing out the relevance of rigorous statistical and probabilistic techniques to pressing software engineering concerns, the panel outlines opportunities for further research in the statistical sciences and their applications to software engineering. The aim is to motivate new researchers from statistics and the mathematical sciences to tackle problems with relevance for software development, as well as to suggest a statistical approach to software engineering concerns that the panel hopes software engineers will find refreshing and stimulating. This report also touches on important issues in training and education for software engineers in the statistical sciences and for statisticians with an interest in software engineering.

Central to this report's theme, and essential to statistical software engineering, is the role of data: *wherever data are used or can be generated in the software life cycle, statistical methods can be brought to bear for description, estimation, and prediction. Nevertheless, the major obstacle to applying statistical methods to software engineering is the lack of consistent, high-quality data* in the resource-allocation, design, review, implementation, and test stages of software development. Statisticians interested in conducting research in software engineering must play a leadership role in justifying that resources are needed to acquire and maintain high-quality and relevant data.

The panel conjectures that the use of adequate metrics and data of good quality is the primary differentiator between successful, productive software development organizations and those that are struggling. Although the single largest area of overlap between statistics and software engineering currently concerns software development and production, it is the panel's view that the largest contributions of statistics to software engineering will be those affecting the quality and productivity of front-end processes, that is, processes that precede code generation. One of the biggest impacts that the statistical community can make in software engineering is to *combine information* across software engineering projects as a means of evaluating effects of technology, language, organization, and process.

CONTENTS OF THIS REPORT

Following an introductory opening chapter intended to familiarize readers with basic statistical software engineering concepts and concerns, a case study of the National Aeronautics and Space Administration (NASA) space shuttle flight control software is presented in Chapter 2 to illustrate some of the statistical issues in software engineering. Chapter 3 describes a well-known general software production model and associated statistical issues and approaches. A critique of some current applications of statistics and software engineering is presented in Chapter 4. Chapter 5 discusses a number of statistical challenges arising in software engineering, and the panel's closing summary and conclusions appear in Chapter 6.

STATISTICAL CHALLENGES

In comparison with other engineering disciplines, software engineering is still in the definition stage. Characteristics of established disciplines include having defined, tested, credible methodologies for practice, assessment, and predictability. Software engineering combines application domain knowledge, computer science, statistics, behavioral science, and human factors issues. Statistical challenges in software engineering discussed in this report include the following:

- Generalizing particular statistical software engineering experimental results to other settings and projects,
- Scaling up results obtained in academic studies to industrial settings,
- Combining information across software engineering projects and studies,
- Adopting exploratory data analysis and visualization techniques,
- Educating the software engineering community regarding statistical approaches and data issues,
- Developing methods of analysis to cope with qualitative variables,

- Providing models with the appropriate error distributions for software engineering applications, and
- Enhancing accelerated life testing.

SUMMARY AND CONCLUSIONS

In the 1990s, complex hardware-based functionality is being replaced by more flexible, software-based functionality, and massive software systems containing millions of lines of code are being created by many programmers with different backgrounds, training, and skills. **The challenge is to build huge, high-quality systems in a cost-effective manner.** The panel expects this challenge to preoccupy the field of software engineering for the rest of the decade. Any set of methodologies that can help in this task will be invaluable. More importantly, the use of such methodologies will likely determine the competitive positions of organizations and nations involved in software production. **What is needed is a detailed understanding by statisticians of the software engineering process, as well as an appreciation by software engineers of what statisticians can and cannot do.**

Catalysts essential for this productive interaction between statisticians and software engineers, and some of the interdisciplinary research opportunities for software engineers and statisticians, include the following:

- **A model for statistical research in software engineering that is collaborative in nature. The ideal collaboration partners statisticians, software engineers, and a *real* software process or product.** Barriers to academic reward and recognition barriers, as well as obstacles to the funding of cross-disciplinary research, can be expected to decrease over time; in the interim, **industry can play a leadership role in nurturing collaborations between software engineers and statisticians** and can reduce its own set of barriers (for instance, those related to proprietary and intellectual property interests).

- **A model for data collection and analysis that ensures the availability of high-quality data for statistical approaches to issues in software engineering.** Careful attention to data issues ranging from definition of metrics to feed-back/-forward loops, including exploratory data analysis, statistical modeling, defect analysis, and so on, is essential if statistical methods are to have any appreciable impact on a given software project under study. For this reason it is crucial that the software industry take a lead position in research on statistical software engineering.

- **Attention to relevant issues in education.** Enormous opportunities and many potential benefits are possible if the software engineering community learns about relevant statistical methods and if statisticians contribute to and cooperate in the education of future software engineers. Some relevant areas include:

—**Designed experiments.** Software engineering is inherently experimental, yet relatively few designed experiments have been conducted. Software engineering education programs must stress the desirability, where feasible, of validating new techniques using statistically valid designed experiments.

—**Exploratory data analysis.** Exploratory data analysis methods are essentially "model free," whereby the investigator hopes to be surprised by *unexpected* behavior rather than having thinking constrained to what is expected.

—**Modeling.** Recent advances in the statistical community in the past decade have effectively relaxed the linearity assumptions of nearly all classical techniques. There should be an emphasis on educational information exchange leading to more and wider use of these recently developed techniques.

—**Risk analysis.** A paradigm for managing risk for the space shuttle program, discussed in Chapter 2 of this report, and the corresponding statistical methods can play a crucial role in identifying risk-prone parts of software systems and of combined hardware and software systems.

—**Attitude toward assumptions.** Software engineers should be aware that violating assumptions is not as important as thoroughly understanding the violation's effects on conclusions. Statistics textbooks, courses, and consulting activities should convey the statistician's level of understanding about and perspective on the importance and implications of assumptions for statistical inference methods.

—**Visualization.** Graphics is important in exploratory stages in helping to ascertain how complex a model the data ought to support; in the analysis stage, by which residuals are displayed to examine what the currently entertained model has failed to account for; and in the presentation stage, in which graphics can provide succinct and convincing summaries of the statistical analysis and the associated uncertainty. Visualization can help software engineers cope with, and understand, the huge quantities of data collected as part of the software development process.

—**Tools.** It is important to identify good statistical computing tools for software engineers. An overview of statistical computing, languages, systems, and packages should be done that is focused specifically for the benefit of software engineers.

1
Introduction

statistics. The mathematics of the collection, organization, and interpretation of numerical data, especially the analysis of population characteristics by inference from sampling.[1]

software engineering. (1) The application of a systematic, disciplined, quantifiable approach to the development, operation, and maintenance of software; that is, the application of engineering to software. (2) The study of approaches as in (1).[2]

statistical software engineering. The interdisciplinary field of statistics and software engineering specializing in the use of statistical methods for controlling and improving the quality and productivity of the practices used in creating software.

The above definitions describe the islands of knowledge and experience that this report attempts to bridge. Software is a critical core industry that is essential to U.S. national interests in science, technology, and defense. It is ubiquitous in today's society, coexisting with hardware (micro-electronic circuitry) in our transportation, communication, financial, and medical systems. The software in a modern cardiac pacemaker, for example, consists of approximately one-half megabyte of code that helps control the pulse rate of patients with heart disorders. In this and other applications, issues such as reliability and safety, fault tolerance, and dependability are obviously important. From the industrial perspective, so also are issues concerned with improving the quality and productivity of the software development process. Yet statistical methods, despite the long history of their impact in manufacturing as well as in traditional areas of science, technology, and medicine, have as yet had little impact on either hardware or software development.

This report is the product of a panel convened by the Board on Mathematical Sciences' Committee on Applied and Theoretical Statistics (CATS) to identify challenges and opportunities in software development and implementation that have a significant statistical component. In attempting to identify interrelated aspects of statistics and software engineering, it enunciates a new interdisciplinary field: *statistical software engineering.* While emphasizing the relevance of applying rigorous statistical and probabilistic techniques to problems in software engineering, the panel also points out opportunities for further research in the statistical sciences and their applications to software engineering. Its hope is that new researchers from statistics and the mathematical sciences will thus be motivated to address relevant and pressing problems of

---

[1] See *The American Heritage Dictionary of the English Language* (1981).
[2] See Institute of Electrical and Electronics Engineers (1990).

software development and also that software engineers will find the statistical emphasis refreshing and stimulating. This report also addresses the important issues of training and education of software engineers in the statistical sciences and of statisticians with an interest in software engineering.

At the panel's information-gathering forum in October 1993, 12 invited speakers described their views on topics that are considered in detail in Chapters 2 through 6 of this report. One of the speakers, John Knight, pointed out that the date of the forum coincided nearly to the day with the 25th anniversary of the Garmisch Conference (Randell and Naur, 1968), a NATO-sponsored workshop at which the term "software engineering" is generally accepted to have originated. The particular irony of this coincidence is that it is also generally accepted that although much more ambitious software systems are now being built, little has changed in the *relative* ability to produce software with predictable quality, costs, and dependability. One of the original Garmisch participants, A.G. Fraser, now associate vice president in the Information Sciences Research Division at AT&T Bell Laboratories, defends the apparent lack of progress by the reminder that prior to Garmisch, there was no "collective realization" that the problems individual organizations were facing were shared across the industry—thus Garmisch was a critical first step toward addressing issues in software production. It is hoped that this report will play a similar role in seeding the field of statistical software engineering by indicating opportunities for statistical thinking to help increase understanding, as well as the productivity and quality, of software and software production.

In preparing this report, the panel struggled with the problem of providing the "big picture" of the software production process, while simultaneously attempting to highlight opportunities for related research on statistical methods. The problems facing the software engineering field are indeed broad, and nonstatistical approaches (e.g., formal methods for verifying program specifications) are at least as relevant as statistical ones. Thus this report tends to emphasize the larger context in which statistical methods must be developed, based on the understanding that recognition of the scope and the boundaries of problems is essential to characterizing the problems and contributing to their solution. It must be noted at the outset, for example, that software engineering is concerned with more than the end product, namely, code. The production process that results in code is a central concern and thus is described in detail in the report. To a large extent, the presentation of material mirrors the steps in the software development process. Although currently the single largest area of overlap between statistics and software engineering concerns software testing (which implies that the code exists), it is the panel's view that the largest contributions to the software engineering field will be those affecting the quality and productivity of the processes that precede code generation.

The panel also emphasizes that the process and methods described in this report pertain to the case of *new* software projects, as well as to the more ordinary circumstance of *evolving* software projects or "legacy systems." For instance, the software that controls the space shuttle flight systems or that runs modern telecommunication networks has been evolving for several decades. These two cases are referred to frequently to illustrate software development concepts and current practice, and although the software systems may be uncharacteristically large, they are arguably forerunners of what lies ahead in many applications. For example, laser printer software is witnessing an order-of-magnitude (base-10) increase in size with each new release.

Similar increases in size and complexity are expected in all consumer electronic products as increased functionality is introduced.

Central to this report's theme, and essential to statistical software engineering, is the role of *data,* the realm where opportunities lie and difficulties begin. The opportunities are clear: *whenever data are used or can be generated in the software life cycle, statistical methods can be brought to bear for description, estimation, and prediction.* This report highlights such areas and gives examples of how statistical methods have been and can be used.

Nevertheless, the major obstacle to applying statistical methods to software engineering is the lack of consistent, high-quality data in the resource-allocation, design, review, implementation, and test stages of software development. Statisticians interested in conducting research in software engineering must acknowledge this fact and play a leadership role in providing adequate grounds for the resources needed to acquire and maintain high-quality, relevant data. A statement by one of the forum participants, David Card, captures the serious problem that statisticians face in demonstrating the value of good data and good data analysis: "It may not be that effective to be able to rigorously demonstrate a 10% or 15% or 20% improvement (in quality or productivity) when with *no* data and *no* analysis, you can claim 50% or even 100%."

The cost of collecting and maintaining high-quality information to support software development is unfortunately high, but arguably essential—as the NASA case study presented in Chapter 2 makes clear. The panel conjectures that use of adequate metrics and data of good quality is, in general, the primary differentiator between successful, productive software development organizations and those that are struggling. Traditional manufacturers have learned the value of investing in an information system to support product development; software development organizations must take heed. All too often, as a release date approaches, all available resources are dedicated to moving a software product out the door, with the result that few or no resources are expended on collecting data during these crucial periods. Subsequent attempts at retrospective analysis—to help forecast costs for a new product or identify root causes of faults found during product testing—are inconclusive when speculation rather than hard data is all that is available to work with. But even software development organizations that realize the importance of historical data can get caught in a downward spiral: effort is expended on collection of data that initially are insufficient to support inferences. When data are not being used, efforts to maintain their quality decrease. But then when the data are needed, their quality is insufficient to allow drawing conclusions. The spiral has begun.

As one means of capturing valuable historical data, efforts are under way to create repositories of data on software development experiments and projects. There is much apprehension in the software engineering community that such data will not be helpful because the relevant metadata (data about the data) are not likely to be included. The panel shares this concern because the exclusion of metadata not only encourages sometimes thoughtless analyses, but also makes it too easy for statisticians to conduct isolated research in software engineering. The panel believes that truly collaborative research must be undertaken and that it must be done with a keen eye to solving the particular problems faced by the software industry. Nevertheless, the panel recognizes benefits to collecting data or experimentation in software development. As is pointed out in more detail in Chapter 5, one of the largest impacts the statistical community can have in software engineering concerns efforts to *combine information* (NRC, 1992) across software engineering projects as a means of evaluating the effects of technology, language, organization, and the development process itself. Although difficult issues are posed by the need to adjust appropriately for differences in projects, the inconsistency of metrics, and varying degrees of data quality, the availability of a data repository at least allows for such research to begin.

Although this report serves as a review of the software production process and related research to date, it is necessarily incomplete. Limitations on the scope of the panel's efforts precluded a fuller treatment of some material and topics as well as inclusion of case studies from a wider variety of business and commercial sectors. The panel resisted the temptation to draw on analogies between software development and the converging area of computer hardware development (which for the most part is initially represented in software). The one approach it is confident of not reflecting is over-simplification of the problem domain itself.

2
Case Study: NASA Space Shuttle Flight Control Software

The National Aeronautics and Space Administration leads the world in research in aeronautics and space-related activities. The space shuttle program, begun in the late 1970s, was designed to support exploration of Earth's atmosphere and to lead the nation back into human exploration of space.

IBM's Federal Systems Division (now Loral), which was contracted to support NASA's shuttle program by developing and maintaining the safety-critical software that controls flight activities, has gained much experience and insight in the development and safe operation of critical software. Throughout the program, the prevailing management philosophy has been that quality must be built into software by using software reliability engineering methodologies. These methodologies are necessarily dependent on the ability to manage, control, measure, and analyze the software using descriptive data collected specifically for tracking and statistical analysis. Based on a presentation by Keller (1993) at the panel's information-gathering forum, the following case study describes space shuttle flight software functionality as well as the software development process that has evolved for the space shuttle program over the past 15 years.

OVERVIEW OF REQUIREMENTS

The primary avionics software system (PASS) is the mission-critical on-board data processing system for NASA's space shuttle fleet. In flight, all shuttle control activities—including main engine throttling, directing control jets to turn the vehicle in a different orientation, firing the engines, or providing guidance commands for landing—are performed manually or automatically with this software. In the event of a PASS failure, there is a backup system. As indicated in the space shuttle flight log history, the backup system has never been invoked.

To ensure high reliability and safety, IBM has designed the space shuttle computer system to have four redundant, synchronized computers, each of which is loaded with an identical version of the PASS. Every 3 to 4 milliseconds, the four computers check with one another to assure that they are in lock step and are doing the same thing, seeing the same input, sending the same output, and so forth. The operating system is designed to instantaneously deselect a failed computer.

The PASS is safety-critical software that must be designed for quality and safety at the outset. It consists of approximately 420,000 lines of source code developed in HAL, an engineering language for real-time systems, and is hosted on flight computers with very limited memory. Software is integrated within the flight control system in the form of overlays—only the small amount of code necessary for a particular phase of the flight (e.g., ascent, on-orbit, or entry activities) is loaded in computer memory at any one time. At quiescent points in the mission, the memory contents are "swapped out" for program applications that are needed for the next phase of the mission.

In support of the development of this safety-critical flight code, there are another 1.4 million lines of code. This additional software is used to build, develop, and test the system as well as to provide simulation capability and perform configuration control. This support software must have the same high quality as the on-board software, given that flawed ground software can mask errors, introduce errors into the flight software, or provide an incorrect configuration of software to be loaded aboard the shuttle.

In short, IBM/Loral maintains approximately 2 million lines of code for NASA's space shuttle flight control system. The continually evolving requirements of NASA's spaceflight program result in an evolving software system: the software for each shuttle mission flown is a composite of code that has been implemented incrementally over 15 years. At any given time, there is a subset of the original code that has never been changed, code that was sequentially added in each update, and new code pertaining to the current release. Approximately 275 people support the space shuttle software development effort.

THE OPERATIONAL LIFE CYCLE

Originally the PASS was developed to provide a basic flight capability of the space shuttle. The first flown version was developed and supported for flights in 1981 through 1982. However, the requirements of the flight missions evolved to include increased operational capability and maintenance flexibility. Among the shuttle program enhancements that changed the flight control system requirements were changes in payload manifest capabilities and main engine control design, crew enhancements, addition of an experimental autopilot for orbiting, system improvements, abort enhancements, provisions for extended landing sites, and hardware platform changes. Following the *Challenger* accident, which was not related to software, many new safety features were added and the software was changed accordingly.

For each release of flight software (called an operational increment), a nominal 6- to 9-month period elapses between delivery to NASA and actual flight. During this time, NASA performs system verification (to assure that the delivered system correctly performs as required) and validation (to assure that the operation is correct for the intended domain). This phase of the software life cycle is critical to assuring safety before a safety-critical operation occurs. It is a time for a complete integrated system test (flight software with flight hardware in operational domain scenarios). Crew training for mission practices is also performed at this time.

A STATISTICAL APPROACH TO MANAGING THE SOFTWARE PRODUCTION PROCESS

To manage the software production process for space shuttle flight control, descriptive data are systematically collected, maintained, and analyzed. At the beginning of the space shuttle program, global measurements were taken to track schedules and costs. But as software development commenced, it became necessary to retain much more product-specific information, owing to the critical nature of space shuttle flight as well as the need for complete accountability for the shuttle's operation. The detail and granularity of data dictate not only the type but also the level of analysis that can be done. Data related to failures have been specifically accumulated in a database along with all the other corollary information available, and a procedure has been established for reliability modeling, statistical analysis, and process improvement based on this information.

A composite description of all space shuttle software of various ages is maintained through a configuration management (CM) system. The CM data include not only a change itself, but also the lines of code affected, reasons for the change, and the date and time of change. In addition, the CM system includes data detailing scenarios for possible failures and the probability of their occurrence, user response procedures, the severity of the failures, the explicit software version and specific lines of code involved, the reasons for no previous detection, how long the fault had existed, and the repair or resolution. Although these data seem abundant, it is important to acknowledge their time dependence, because the software system they describe is subject to constant "churn."

Over the years, the CM system for the space shuttle program has evolved into a common, minimum set of data that must be retained regarding every fault that is recognized anywhere in the life cycle, including faults found by inspections before software is actually built. This evolutionary development is amenable to evaluation by statistical methods. Trend analysis and predictions regarding testing, allocation of resources, and estimation of probabilities of failure are examples of the many activities that draw on the database. This database also continues to be the basis for defining and developing sophisticated, insightful estimation techniques such as those described by Munson (1993).

Fault Detection

Management philosophy prescribes that process improvement is part of the process. Such proactive process improvement includes inspection at every step of the process, detailed documentation of the process, and analysis of the process itself.

The critical implications of an ill-timed failure in space shuttle flight control software require that remedies be decisive and aggressive. When a fault is identified, a feedback process involving detailed information on the fault enforces a search for similar faults in the existing system and changes the process to guard actively against such faults in flight control software development. The characteristics of a single fault are actively documented in the following four-step reactive process-improvement protocol:

1. Remove the fault,
2. Identify the root cause of the fault,
3. Eliminate the process deficiency that let the fault escape earlier detection, and
4. Analyze the product for other, similar faults.

Further scrutiny of what occurred in the process between introduction and detection of a fault is aimed at determining why downstream process elements failed to detect and remove the fault. Such introspective analysis is designed to improve the process and specific process elements so that if a similar fault is introduced again, these process elements will detect it before it gets too far along in the product life cycle. This four-step process improvement is achievable because of the maturity of the overall IBM/Loral software management process. The complete recording of project events in the CM system (phase of the process, change history of involved line(s) of code, the line of code that included an error, the individuals involved, and so on) allows hindsight so that the development team can approach the occurrence of an error not as a failure but rather as an opportunity to improve the process and to find other, similar errors.

Safety Certification

The dependability of safety-critical software cannot be based merely on testing the software, counting and repairing the faults, and conducting "live tests" on shuttle missions. Testing of software for many, many years, much longer than its life cycle, would be required in order to demonstrate software failure probability levels of 10^{-7} or 10^{-9} per operational hour. A process must be established, and it must be demonstrated statistically that if that process is followed and maintained under statistical control, then software of known quality will result. One result is the ability to predict a particular level of fault density, in the sense that fault density is proportional to failure intensity, and so provide a confidence level regarding software quality. This approach is designed to ensure that quality is built into the software at a measurable level. IBM's historical data demonstrate a constantly improving process for comfort of space shuttle flight. The use of software engineering methodologies that incorporate statistical analysis methods generally allows the establishment of a benchmark for obtaining a valid measure of how well a product meets a specified level of quality.

3
A Software Production Model

The software development process spans the life cycle of a given project, from the first idea, to implementation, through completion. Many process models found in the literature describe what is basically a problem-solving effort. The one discussed in detail below, as a convenient way to organize the presentation, is often described as the waterfall model. It is the basis for nearly all the major software products in use today. But as with all great workhorses, it is beginning to show its age. New models in current use include those with design and implementation occurring in parallel (e.g., rapid prototyping environments) and those adopting a more integrated, less linear, view of a process (e.g., the spiral model referred to in Chapter 6). Although the discussion in this chapter is specific to a particular model, that in subsequent chapters cuts across all models and emphasizes the need to incorporate statistical insight into the measurement, data collection, and analysis aspects of software production.

The first step of the software life cycle (Boehm, 1981) is the generation of system requirements whereby functionality, interactions, and performance of the software product are specified in (usually) numerous documents. In the design step, system requirements are refined into a complete product design, an overall hardware and software architecture, and detailed descriptions of the system control, data, and interfaces. The result of the design step is (usually) a set of documents laying out the system's structure in sufficient detail to ensure that the software will meet system requirements. Most often, both requirements and design documents are formally reviewed prior to coding in order to avoid errors caused by incorrectly stated requirements or poor design. The coding stage commences once these reviews are successfully completed. Sometimes scheduling considerations lead to parallel review and coding activities. Normally individuals or small teams are assigned specific modules to code. Code inspections help ensure that module quality, functionality, and schedule are maintained.

Once modules are coded, the testing step begins. (This topic is discussed in some detail in Chapter 3.) Testing is done incrementally on individual modules (unit testing), on sets of modules (integration testing), and finally on all modules (system testing). Inevitably, faults are uncovered in testing and are formally documented as modification requests (MRs). Once all MRs are resolved, or more usually as schedules dictate, the software is released. Field experience is relayed back to the developer as the software is "burned in" in a production environment. Patches or rereleases follow based on customer response. Backward compatibility tests (regression testing) are conducted to ensure that correct functionality is maintained when new versions of the software are produced.

The above overview is noticeably nonquantitative. Indeed, this nonquantitative characteristic is the most striking difference between software engineering and more traditional (hardware) engineering disciplines. Measurement of software is critical for characterizing both the process and the product, and yet such measurement has proven to be elusive and controversial. As argued in Chapter 1, the application of statistical methods is predicated on the existence of relevant data, and the issue of software measurements and metrics is discussed prominently throughout the report. This is not to imply that measurements have never been made or that data are totally lacking. Unfortunately metrics tend to describe properties and conditions for which it is easy to gather data rather than those that are useful for characterizing software content, complexity, and form.

PROBLEM FORMULATION AND SPECIFICATION OF REQUIREMENTS

Within the context of system development, specifications for required software functions are derived from the larger system requirements, which are the primary source for determining what the delivered software product will do and how it will do it. These requirements are translated by the designer or design team into a finished product that delivers all that is explicitly stated and does not include anything explicitly forbidden. Some common references regarding requirements specification are mentioned in *IEEE Standard for Software Productivity Metrics* (IEEE, 1993).

Requirements—the first formal tangible product obtained in the development of a system—are subjective statements specifying the system's various desired operational characteristics. Errors in requirements arise for a number of reasons, including ambiguous statements, inconsistent information, unclear user requirements, and incomplete requests. Projects that have ill-defined or unstated requirements are subject to constant iteration, and a lack of precise requirements is a key source of subsequent software faults. In general, the longer a fault resides in a system before it is detected, the greater is the cost of removing it or recovering from related failures. This condition is a primary driver of the review process throughout software development.

The formulation requirements start with customers requesting a new functionality. Systems engineers collect information describing the new functionality and develop a customer specification description (CSD) describing the customer's view of the feature. The CSD is used internally by software development organizations to formulate cost estimates for bidding. After the feature is committed (sold), systems engineers write a feature specification description (FSD) describing the internal view of the feature. The FSD is commonly referred to as "requirements." Both the CSD and FSD are carefully reviewed and must meet formal criteria for approval.

DESIGN

The heart of the software development cycle is the translation and refinement of the requirements into code. Software architects transform the requirements for each specified feature into a high-level design. As part of this process, they determine which subsystems (e.g., databases) and modules are required and how they interact or communicate. The broad, high-level design is then refined into a detailed low-level design. This transformation involves much information gathering and detective work. The software architects are often the most experienced and knowledgeable of the software engineers.

The sequence of continual refinements ultimately results in a mapping of high-level functions into modules and code. Part of this design process is selecting an appropriate representation, which in most cases is a specific programming language. Selection of a representation involves factors such as operational domain, system performance, and function, among others. When completed, the high-level design is reviewed by all, including those concerned with the affected subsystems and the organization responsible for development.

The human element is a critical issue in the early stages of a software project. Quantitative data are potentially available following document reviews. Specifically, early in the development cycle of software systems, (paper) documents are prepared describing feature requirements or feature design. Prior to a formal document review, the reviewers individually read the document, noting issues that they believe should be resolved before the document is approved and feature development is begun. At the review meeting, a single list of issues is prepared that includes the issues noted by the reviewers as well as the ones discovered during the meeting itself. This process thus generates data consisting of a tabulation of issues found by each reviewer. The degree of overlap provides information regarding the number of remaining issues, that is, those yet to be identified. If this number is acceptably small, the process can proceed to the next step; if not, further document refinement is necessary in order to avoid costly fixes later in the process. The problem as stated bears a certain resemblance to capture-recapture models in wildlife studies, and so appropriate statistical methods can be devised for analyzing the review data, as illustrated in the following example.

Example. Table 1 contains data on issues identified for a particular feature for the AT&T 5ESS switch (Eick et al., 1992a). Six reviewers found a total of 47 distinct issues. A common capture-recapture model assumes that each issue has the same probability of being captured (detected) and that reviewers work independently with their own chance of capturing an issue, or detection probability. Under such a model, likelihood methods yield an estimate of $N = 65$, implying that approximately 20 issues remain to be identified in the document. An upper 95% confidence bound for N under this model is 94 issues.

Such a model is natural but simplistic. The software development environment is not conducive to independence among reviewers (so that some degree of collusion is unavoidable), and reviewers also are selected to cover certain areas of specialization. In either case, the cornerstone of capture-recapture models, the binomial distribution, is no longer appropriate for the total number of issues. It is possible to develop a likelihood-based test for pairwise collusion of reviewers and reviewer-specific tests of specialization. In the example above, there is no evidence of collusion among reviewers, but reviewer C exhibits a significantly greater degree of specialization than do the other reviewers. When this reviewer is treated as a specialist, the maximum likelihood estimate (MLE) of the number of issues is reduced to 53, implying that only a half dozen issues remain to be discovered in the document. ∎

Other mismatches between the data arising in software review and those in capture-recapture wildlife population studies induce bias in the MLE. Another possible estimator for this problem is the jackknife estimator (Burnham and Overton, 1978). But this estimator seems in fact to be more biased than the MLE (Vander Wiel and Votta, 1993). Both are rescued to a large extent by their categorization of faults into classes (e.g., "easy to find" versus "hard to find"). In any given

Table 1. Issue discovery. The rows of the table represent 47 issues noted by six reviewers prior to review meetings. An entry in cell i,j of the table indicates that issue i (i = 1,...,47) was noted by reviewer j (j = A,...,F). Rows with no entries (i.e., column sums of zero) correspond to issues discovered at the meeting.

Issue	A	B	C	D	E	F	Sum
1	1						1
2				1		1	2
3		\	1				1
4	1						1
5							0
6				1			1
7				1			1
8				1			1
9							0
10	1						1
11	1						1
12				1			1
13							0
14	1			1			2
15	1						1
16							0
17	1			1			2
18	1						1
19	1			1			2
20	1			1			2
21	1	1		1	1	1	5
22	1					1	2
23		1					1
24						1	1

Issue	A	B	C	D	E	F	Sum
25	1			1			2
26	1			1			2
27					1		1
28				1			1
29				1	1		2
30	1						1
31					1		1
32			1				1
33	1						1
34	1			1	1		3
35	1			1			2
36	1						1
37	1						1
38			1				1
39			1				1
40						1	1
41			1				1
42	1					1	2
43	1						1
44	1						1
45					1		1
46	1						1
47	1						1
SUM	25	3	4	13	9	6	60

application, it is necessary to verify that the "easy to find" and "hard to find" classification is meaningful, or to determine that it is merely partitioning the distribution of difficulty in an arbitrary manner. A relevant point in this and other applications of statistical methods in software engineering is that addressing aspects of the problem that induce study bias is important and valued—theoretical work addressing aspects of statistical bias is not likely to be as highly valued.

IMPLEMENTATION

The phase in the software development process that is often referred to interchangeably as coding, development, or implementation is the actual transformation of the requirements into executable form. "Implementation in the small" refers to coding, and "implementation in the large" refers to designing an entire system in a top-down fashion while maintaining a perspective on the final integrated system.

Low-level designs, or coding units, are created from the high-level design for each subsystem and module that needs to be changed. Each coding unit specifies the changes to be made to the existing files, new or modified entry points, and any file that must be added, as well as other changes. After document reviews and approvals, the coding may begin. Using private copies of the code, developers make the changes and add the files specified in the coding unit. Coding is delicate work, and great care is taken so that unwanted side effects do not break any of the existing code. After completion, the code is tested by the developer and carefully reviewed by other experts. The changes are submitted to a public load (code from all programmers that is merged and loaded simultaneously) using an MR number. The MR is tied back to the feature to establish a correspondence between the code and the functionality that it provides.

MRs are associated with the system version management system, which maintains a complete history of every change to the software and can recreate the code as it existed at any point in time. For production software systems, version management systems are required to ensure code integrity, to support multiple simultaneous releases, and to facilitate maintenance. If there is a problem, it may be necessary to back out changes. Besides a record of the affected lines, other information is kept, such as the name of the programmer making the changes, the associated feature number, whether a change fixes a fault or adds new functionality, the date of a change, and so on.

The configuration management database contains the record of code changes, or change history of the code. Eick et al. (1992b) describe a visualization technique for displaying the change history of source code. The graphical technique represents each file as a vertical column and each line of code as a color-coded row within the column. The row indentation and length track the corresponding text, and the row color is tied to a statistic. If the row tracking is literal as with computer source code, the display looks as if the text had been printed in color and then photo-reduced for viewing as a single figure. The spatial pattern of color shows the distribution of the statistic within the text.

Example. Developing large software systems is a problem of scale. In multimillion-line systems there may be hundreds of thousands of files and tens of thousands of modules, worked on by thousands of programmers for multiyear periods. Just discovering what the existing code does is a major technical problem consuming significant amounts of time. A continuing and significant problem is that of code discovery, whereby programmers try to understand how unfamiliar code works. It may take several weeks of detailed study to change a few lines of code without causing unwanted side effects. Indeed, much of the effort in maintenance involves changing code written by another programmer. Because of variation in programmer staff sizes and inevitable turnover, training new programmers is important. Visualization techniques, described further in Chapter 5, can improve productivity dramatically.

Figure 1 displays a module composed of 20 source code files containing 9,365 lines of code. The height of each column indicates the size of the file. Files longer than one column are continued over to the next. The row color indicates the age of each line of code using a rainbow color scale with the newest lines in red and the oldest in blue. On the left is an interactive color scale showing a color for each of the 324 changes by the 126 programmers modifying this code over the last 10 years. The visual impression is that of a miniature picture of all of the source code, with the indentation showing the usual C language control structure.

The perception of colors is blurred, but there are clear patterns. Files in approximately the same hue were written at about the same time and are related. Rainbow files with many different hues are unstable and are likely to be trouble spots because of all the changes. The biggest file has about 1,300 lines of code and takes a column and a half. ∎

Changes from many coding units are periodically combined together into a so-called common load of the software system. The load is compiled, made available to developers for testing, and installed in the laboratory machines. Bringing the changes together is necessary so that developers working on different coding units of a common feature can ensure that their code works together properly and does not break any other functionality. Developers also use the public load to test their code on laboratory machines.

After all coding units associated with a feature are complete and it has been tested by the developers in the laboratory, the feature is turned over to the integration group for independent testing. The integration group runs tests of the feature according to a feature test plan that was prepared in parallel with the FSD. Eventually the new code is released as part of an upgrade or sent out directly if it fixes a critical fault. At this stage, maintenance on the code begins. If customers have problems, developers will need to submit fault modification requests.

TESTING

Many software systems in use today are very large. For example, the software that supports modern telecommunications networks, or processes banking transactions, or checks individual tax returns for the Internal Revenue Service has millions of lines of code. The development of such large-scale software systems is a complex and expensive process. Because a single simple fault in a system may cripple the whole system and result in a significant loss (e.g., loss of telephone service in an entire city), great care is needed to assure that the system is flawlessly constructed. Because a fault can occur in only a small part of a system, it is necessary to assure that even small programs are working as intended. Such checking for conformance is accomplished by testing the software.

Specifically, the purpose of software testing is to detect errors in a program and, in the absence of errors, gain confidence in the correctness of the program or the system under test. Although testing is no substitute for improving a process, it does play a crucial role in the overall software development process. Testing is important because it is effective, if costly. It is variously estimated that the total cost of testing is approximately 20 to 33% of the total software budget for software development (Humphrey, 1989). This fraction amounts to billions of dollars in the U.S. software industry alone. Further, software testing is very time consuming, because the time for testing is typically greater than that for coding. Thus, efforts to reduce the costs and improve the effectiveness of testing can yield substantial gains in software quality and productivity.

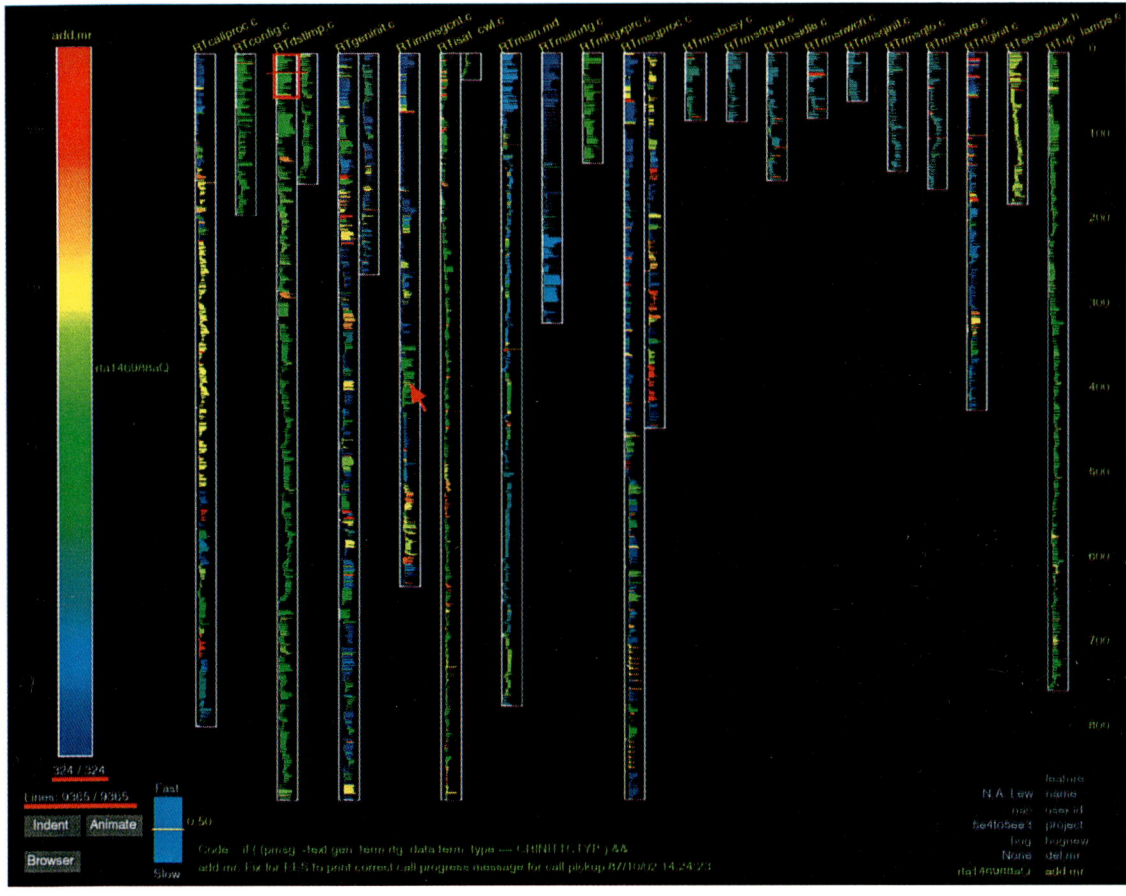

Figure 1. A *SeeSoft*[TM] display showing a module with 20 files and 9,365 lines of code. Each file is represented as a column and each line of code as a colored row. The newest rows are in red and the oldest in blue, with a color spectrum in between. This overview highlights the largest files and program control structures, while the color shows relationships between files, as well as unstable, frequently changed code. Source: Eick *et al.* (1992b).

Much of the difficulty of software testing is in the management of the testing process (producing reports, entering MRs, documenting MRs cleared, and so on), the management of the objects of the testing process (test cases, test drivers, scripts, and so on), and the management of the costs and time of testing.

Typically, software testing refers to the phase of testing carried out after parts of code are written so that individual programs or modules can be compiled. This phase includes unit, integration, system, product, customer, and regression testing. Unit testing occurs when programmers test their own programs, and integration testing is the testing of previously separate parts of the software when they are put together. System testing is the testing of a functional part of the software to determine whether it performs its expected function. Product testing is meant to test the functionality of the final system. Customer testing is often product testing performed by the intended user of the system. Regression testing is meant to assure that a new version of a system faithfully reproduces the desirable behavior of the previous system.

Besides the stages of testing, there are many different testing methods. In white box testing, tests are designed on the basis of detailed architectural knowledge of the software under test. In black box testing, only knowledge of the functionality of the software is used for testing; knowledge of the detailed architectural structure or of the procedures used in coding is not used. White box testing is typically used during unit testing, in which the tester (who is usually the developer who created the code) knows the internal structure and tries to exercise it based on detailed knowledge of the code. Black box testing is used during integration and system testing, which emphasizes the user perspective more than the internal workings of the software. Thus, black box testing tries to test the functionality of software by subjecting the system under test to various user-controlled inputs and by assessing its resulting performance and behavior.

Since the number of possible inputs or test cases is almost limitless, testers need to select a sample, a suite of test cases, based on their effectiveness and adequacy. Herein lie significant opportunities for statistical approaches, especially as applied to black box testing. Ad hoc black box testing can be done when testers, perhaps based on their knowledge of the system under test and its users, decide specific inputs. Another approach, based on statistical sampling ideas, is to generate test cases randomly. The results of this testing can be analyzed by using various types of reliability growth curve models (see "Assessment and Reliability" in Chapter 4). Random generation requires a statistical distribution. Since the purpose of black box testing is to simulate actual usage, a highly recommended technique is to generate test cases randomly from the statistical distribution needed by users, often referred to as the operational profile of a system.

There are several advantages and disadvantages to statistical operational profile testing. A key advantage is that if one takes a large enough sample, then the system under test will be tested in all the ways that a user may need it and thus should experience fewer field faults. Another advantage of this method is the possibility of bringing the full force of statistical techniques to bear on inferential problems; that is, the results obtained during testing can be generalized to make inferences about the field behavior of the system under test, including inferences about the number of faults remaining, the failure rate in the field, and so on.

In spite of all these advantages, statistical operational profile testing in its purest form is rarely used. There are many difficulties; some are operational and others are more basic. For example, one can never be certain about the operational profile in terms of inputs, and especially in terms of their probabilities of occurrence. Also, for large systems, the input space is high-dimensional. Thus, another problem is how to sample from this high-dimensional space. Further, the distribution is not static; it will, in all likelihood, change over time as new users exercise the system in unanticipated ways. Even if this possibility can be discounted, questions remain about the efficiency of statistical operational profile testing, which can be very inefficient, because most often the system under test will be used in routine ways, and thus a randomly drawn sample will be highly weighted by routine operations. This high weighting may be fine if the number of test cases is very large. But then testing would be very expensive, perhaps even prohibitively so. Therefore, testers often adopt some variant of drawing a random sample; for example, testers give more weight to boundary values—those values around which the system is expected to change its behavior and therefore where faults are likely to be found. This and other clever strategies adopted by testers typically result in a testing distribution that is quite different from the operational profile. Of course, in such a case the results of the testing laboratory will not be generalizable unless the relationships between the two distributions are taken into account.

Thus, to take advantage of the attractiveness of operational profile testing, some key problems have to be solved:

1. How to obtain the operational profile,
2. How to sample according to a statistical distribution in high-dimensional space, and
3. How to generalize results obtained in the testing laboratory to the field when the testing distribution is a variant of the operational profile distribution.

All of these questions can be dealt with conceptually using statistical approaches.

For (1), a Bayesian elicitation procedure can be envisioned to derive the operational profile. This elicitation is done routinely in Bayesian applications, but because the space is very high dimensional, techniques are needed for Bayesian elicitation in very high dimensional spaces.

Concerning (2), if the joint distribution corresponding to the operational profile is known, schemes can be used that are more efficient than simple random sampling schemes. Simple random sampling is inefficient because it typically gives higher probability to the middle of a distribution than to its tails, especially in high dimensions. A more efficient scheme would sample the tails quickly. This can be accomplished by stratifying the support of the distribution.

McKay et al. (1979) formalized this idea using Latin hyper cube sampling. Suppose we have a K-dimensional random vector $X = (X_1,...,X_K)$ and we want to get a sample of size N from the joint distribution of X. If the components of X are independent, then the scheme is simple, namely:

- Divide the range of each component random variable in N intervals of equal probability,
- Randomly sample one observation for each component random variable in each of the corresponding N intervals, and finally
- Randomly combine the components to create X.

Stein (1987) showed that this sampling scheme can be substantially better than simple random sampling. Iman and Conover (1982) and Stein (1987) both discussed extensions for nonindependent component variables. Of course, if specifying homogenous strata is possible, it should be done prior to applying the Latin hyper cube sampling method to increase the overall effectiveness of the sampling scheme.

Example: Consider a software system controlling the state of an air-to-ground missile. The key inputs for the software are *altitude, attack* and *bank* angles, *speed, pitch, roll,* and *yaw*. Typically, these variables are independently controlled. To test this software system, combinations of all these inputs must be provided and the output from the software system checked against the corresponding physics. One would like to generate test cases that include inputs over a broad range of permissible values. To test all the valid possibilities, it would be reasonable to try uniform distributions for each input. Suppose we decide upon a sample of size 6. The corresponding Latin hyper cube design is easily constructed by dividing each variable into six equal probability intervals and sampling randomly from each interval. Because we have independent random variables here, the final step consists of randomly *coupling* these samples. The design is difficult to visualize in more than two dimensions, but one such sample for *attack* and *bank* angles is depicted in Figure 2. Note that there is exactly one observation in each column and in each row, thus the name "Latin hyper cube." ∎

Figure 2. Latin hyper cube. N = 6 and K = 2.

Finally, concerning (3), to make inferences about field performance, the issue of the discrepancy between the statistical operational profile and the testing distribution must be addressed. At this point, a distinction can be made between two types of extrapolation to field performance of the system under test. It is clear that even if the true operational profile distribution is not available, to the extent that the testing distribution has the same support as the operational profile distribution, statistical inferences can be made about the number of remaining faults. On the other hand, to extrapolate the failure intensity from the testing laboratory to the field, it is not enough to have the same support; rather, identical distributions are needed. Of course, it is unlikely that after spending much time and money on testing, one would again test with the statistical operational profile. What is needed is a way of reusing the information generated in the testing laboratory, perhaps by a transformation in which some statistical techniques based on reweighting can help. There are two basic ideas, both relying heavily on the assumption that the testing and the field-use distributions have the same support. One idea is to use all the data from the testing laboratory, but with added weights to change the sample to resemble a random sample from the operational profile. The approach is similar to reweighting in importance sampling. Another idea is to accept or reject the inputs used in testing with a probability distribution based on the operational profile. For a description of both of these techniques, see Beckman and McKay (1987).

In his presentation at the panel's forum, Phadke (1993) suggested another set of statistical techniques, based on orthogonal arrays, for parsimonious testing of software. The example described above proves useful in an elaboration.

Example. For the software system that determines the state of an attack plane, let us assume that interest centers on testing only two conditions for each input variable. This situation arises, for example, when the primary interest lies in boundary value testing. Let the lower value be input state 0 and the upper value be input state 1 for each of the variables. Then in the language of statistical experimental design, we have seven factors, A,...,G (altitude, attack angle, bank angle, speed, pitch, roll, and yaw), each at two levels (0,1). To test all of the possible combinations, one would need a complete factorial experiment, which would have $2^7 = 128$ test cases consisting of all possible sequences of 0's and 1's. For a statistical experiment intended to address only main effects, a highly fractionated factorial design would be sufficient. However, in the case of software testing, there is no statistical variability and little or no interest in estimating various effects. Rather, the interest is in covering the test space as much as possible and checking whether the test cases pass or fail. Even in this case, it is still possible to use statistical design ideas. For example, consider the sequence of test cases given in Table 2. This design requires 8 test cases instead of 128. In this case, since there is no statistical variation, main effects do not have any practical meaning. However, looking at the pattern in the table, it is clear that all possible combinations of any two pairs are covered in a balanced way. Thus, testing according to this design will protect against any incorrect implementation of the code involving a pairwise interaction. ∎

	A	B	C	D	E	F	G
1	0	0	0	0	0	0	0
2	0	0	0	1	1	1	1
3	0	1	1	0	0	1	1
4	0	1	1	1	1	0	0
5	1	0	1	0	1	0	1
6	1	0	1	1	0	1	0
7	1	1	0	0	1	1	0
8	1	1	0	1	0	0	1

Table 2a. Orthogonal array. Test cases in rows. Test factors in columns.

	A	B	C	D	E	F	G
1	0	0	0	0	0	0	0
2	1	1	1	1	1	1	0
3	0	0	1	1	1	0	1
4	1	0	0	0	1	1	1
5	1	1	1	0	0	0	1
6	0	1	0	1	0	1	1

Table 2b. Combinatorial design. Test cases in rows. Test factors in columns.

In general, following Taguchi, Phadke (1993) suggests orthogonal array designs of strength two. These designs (a specific instance of which is given in the above example) guarantee that all possible pairwise combinations will be tried out in a balanced way. Another approach based on combinatorial designs was proposed by Cohen et al. (1994). Their designs do not consider balance to be an overriding design criterion, and accordingly they produce designs with smaller numbers of required test cases. For example, Table 2b contains a combinatorial design with complete pairwise coverage in six runs instead of the eight required by orthogonal arrays (Table 2a). This notion has been extended to notions of higher-order coverage as well. The efficacy of these and other types of designs has to be evaluated in the testing context.

Besides the types of testing discussed above, there are other statistical strategies that can be used. For example, DeMillo et al. (1988) have suggested the use of fault insertion techniques. The basic idea is akin to capture-recapture sampling in which sampled units of a population (usually wildlife) are released and inverse sampling is done to estimate the unknown population size. The MOTHRA system built by DeMillo and his colleagues implements such a scheme. While there are many possible sampling schemes (Nayak, 1988), the difficulty with fault insertion is that the faults inserted ought to be subtle enough so that the system can be compiled and tested; no two inserted faults should interact with each other; and while it may be possible at the unit testing level, it is prohibitively expensive for integration testing. It should be pointed out that the use of capture-recapture sampling, outlined in this chapter's subsection titled "Design," for quantifying document reviews does not require fault seeding and, accordingly, is not subject to the above difficulties.

Another key problem in testing is determining when there has been enough testing. For unit testing where much of the testing is white box and the modules are small, one can attempt to check whether all the paths have been covered by the test cases, an idea extended substantially by Horgan and London (1992). However, for integration and system testing, this particular approach, coverage testing, is not possible because of the size and the number of possible paths through the system. Here is another opportunity for using statistical approaches to develop a theory of statistical coverage. Coverage testing relates to deriving methods and algorithms for generating test cases so that one can state, with a very high probability, that one has checked most of the important paths of the software. This kind of methodology has been used with probabilistic algorithms in protocol testing, where the structure of the program can be described in great detail. (A protocol is a very precise description of the interface between two diverse systems.) Lee and Yanakakis (1992) have proposed algorithms whereby one is guaranteed, with a high degree of probability, that all the states of the protocols are checked. The difficulty with this approach is that the number of states becomes large very quickly, and except for a small part of the system under test, it is not clear that such a technique would be practical (under current computing technology). These ideas have been mathematically formalized in the vibrant area of theorem checking and proving (Blum et al., 1990). The key idea is to take transforms of programs such that the results are invariant under these transforms if the software is correct. Thus, any variation in the results suggests possible faults in the software. Blum et al. (1989) and Lipton (1989), among others, have developed a number of algorithms to give probabilistic bounds on the correctness of software based on the number of different transformations.

In all of the several approaches to testing discussed above, the number of test cases can be extraordinarily large. Because of the cost of testing and the need to supply software in a reasonable period of time, it is necessary to formulate rules about when to stop testing. Herein lies another set of interesting problems in sequential analysis and statistical decision theory. As pointed out by Dalal and Mallows (1988, 1990, 1992), Singpurwalla (1991), and others, the key issue is to explicitly incorporate the economic trade-off between the decision to stop testing (and absorb the cost of fixing subsequent field faults) and the decision to continue testing (and incur ongoing costs to find and fix faults before release of a software product). Since the testing process is not deterministic, the fault-finding process is modeled by a stochastic reliability model (see Chapter 4 for further discussion). The opportune moment for release is decided using sequential decision theory. The rules are simple to implement and have been used in a number of projects. This framework has been extended to the problem of buying software with some sort of probabilistic guarantee on the number of faults remaining (Dalal and Mallows, 1992). Another extension with practical importance (Dalal and McIntosh, 1994) deals with the issue of a system under test not having been completely delivered at the start of testing. This situation is a common occurrence for large systems, where in order to meet scheduling milestones, testing begins immediately on modules and sets of modules as they are completed.

4
Critique of Some Current Applications of Statistics in Software Engineering

COST ESTIMATION

One of software engineering's long-standing problems is the considerable inaccuracy of the cost, resource, and schedule estimates developed for projects. These estimates often differ from the final costs by a factor of two or more. Such inaccuracies have a severe impact on process integrity and ultimately on final software quality. Five factors contribute to this continuing problem:

1. Most cost estimates have little statistical basis and have not been validated;

2. The value of historical data in developing predictive models is limited, since no consistent software development process has been adopted by an organization;

3. The maturity of an organization's process changes the granularity of the data that can be used effectively in project cost estimation;

4. The reliability of inputs to cost estimation models varies widely; and

5. Managers attempt to manage to the estimates, reducing the validity of historical data as a basis for validation.

Certain of the above issues center on the so-called maturity of an organization (Humphrey, 1988). From a purely statistical research perspective, (5) may be the most interesting area, but the major challenge facing the software community is finding the right metrics to measure in the first place.

Example. The data plotted in Figure 3 pertain to the productivity of a conventional COBOL development environment (Kitchenham, 1992). For each of 46 different products, size (number of entities and transactions) and effort (in person-hours) were measured. From Figure 3, it is apparent that despite substantial variability, a strong (log-log) linear relationship exists between program size and program effort.

A simple model relating effort to size is

$$\log_{10}(\mathit{effort}) = \alpha + \beta \log_{10}(\mathit{size}) + \mathit{noise}.$$

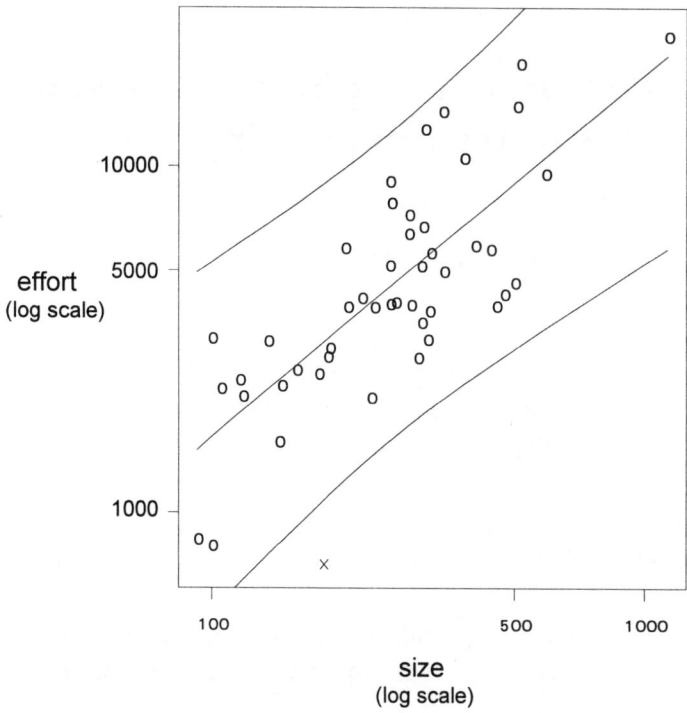

Figure 3. Data on the relationship between development effort and product size in a COBOL development organization.

A least squares fit to these data yields

	Coeff.	SE	t
Intercept	1.120	0.3024	3.702
$\log_{10}(\text{size})$	1.049	0.1250	8.397
RMS	0.194		

These fitted coefficients suggest that development effort is proportional to product size; a formal test of the hypothesis, H: $\beta = 1$, gives a t value at the .65 significance level.

The estimated intercept after fixing $\beta = 1$ is 1.24; the resulting fit and a 95% prediction interval are overlaid on the data in Figure 3. This model predicts that it requires approximately 17 hours ($= 10^{1.24}$) to implement each unit of size.

Such models are used for prediction and tool validation. Consider an additional observation made of a product developed using a fourth-generation language and relational databases. Under the experimental development process, it took 710 hours to implement the product of size 183 (this point is denoted by X in Figure 3). The fitted model predicts that this product would have taken approximately 3,000 hours to complete using the conventional development environment. The 95% prediction interval at X = 183 ranges from approximately 1,000 to 9,000 hours; thus, assuming that other factors are not contributing to the apparent short development cycle of this product, the use of those new fourth-generation tools has demonstrably decreased the development effort (and hence the cost). ∎

Statistical Inadequacies in Estimating

Most cost estimation methods develop an initial relationship between the estimated size of a system (in lines of code, for instance) and the resources required to develop it. Such equations are often of the form illustrated in the above example: effort is proportional to size raised to the β power. This initial estimate is then adjusted by a number of factors that are thought to affect the productivity of the specific project, such as the experience of the assigned staff, the available tools, the requirements for reliability, and the complexity of the interaction with the customer. Thus the estimating equation assumes the log linear form:

$$effort \approx \alpha \; size^{\beta} \times a_i a_j a_k a_l a_m \cdots a_z,$$

where the a's are the coefficients for the adjustment factors. Unfortunately, these adjustment factors are not treated as variables in a regression equation; rather, each has a set of fixed coefficients (termed "weighting factors") associated with each level of the variable. These are independently applied as if the variables were uncorrelated (an assumption known to be incorrect). These weighting schemes have been developed based on intuition about each variable's potential impact rather than on a statistical model fitting using historical data. Thus, although the relationship between effort and size is often recalibrated for different organizations, the weighting factors are not.

Exacerbating the problems with existing cost estimation models is the lack of rigorous validation of the equations. For instance, Boehm (1981) has acknowledged that his well-known COCOMO estimating model was not developed using statistical methods. Many individuals marketing cost estimation modeling tools denigrate the value of statistical approaches compared to clever intuition. To the extent that analytical methods are used in the development or validation of these models, they are often performed on data sets that contain as many predictor variables (productivity factors) as projects. Thus determination of the separate or individual contributions of the variables almost certainly depends too much on chance and can be distorted by collinear relationships. These models are rarely subjected to independent validation studies. Further, little research has been done that attempts to restrict these models to including only those productivity factors that really matter (i.e., subset selection).

Because of the lack of statistical rigor in most cost estimation models, software development organizations usually handcraft weighting schemes to fit their historical results. Thus, the specific instantiation of most cost estimation models differs across organizations. Under these conditions, cross-validation of the weighting schemes is very difficult, if not impossible. A new approach to developing cost estimation models would be beneficial, one that invokes sound statistical principles in fitting such equations to historical data and to validating their applicability across organizations. If the instantiation of such models is found to be domain-specific, statistically valid methods should be sought for regenerating accurate models in different domains.

Process Volatility

In immature software development organizations, the processes used differ across projects because they are based on the experiences and preferences of the individuals assigned to each project, rather than on common organizational practice. Thus, in such organizations cost estimation models must attempt to predict the results of a process that varies widely across projects. In poorly run projects the signal-to-noise ratio is low, in that there is little consistent practice that can be used as the basis for dependable prediction. In such projects, neither the size nor the productivity factors provide any consistent insight into the resources required, since they are not systematically related to the processes that will be used.

The historical data collected from projects in immature software development organizations are difficult to interpret because they reflect widely divergent practices. Such data sets do not provide an adequate basis for validation, since process variation can mask underlying relationships. In fact, because the relationships among independent variables may change with variations in the process, different projects may require different values of the parameters in the cost estimation models. As organizations mature and stabilize their processes, the accuracy of the estimating models they use usually increases.

Maturity and Data Granularity

In mature organizations the software development process is well defined and is applied consistently across projects. The more carefully defined the process, the finer the granularity of the processes that can be measured. Thus, as software organizations mature, the entire basis for their cost estimation models can change. Immature organizations have data only at the level of overall project size, number of person-years required, and overall cost. With increasing organizational maturity, it becomes possible to obtain data on process details such as how many reviews must be conducted at each life cycle stage based on the size of the system, how many test cases must be run, and how many defects must be fixed based on the defect removal efficiency of each stage of the verification process. Thus, estimation in fully developed organizations can be based on a bottom-up analysis in which the historical data can be more accurate because the objects of estimation, and the effort they require, are more easily characterized.

As organizations mature, the structure of relevant cost estimation models can change. When process models are not defined in detail, models must take the form of regression equations based on variables that describe the total impact of a predictor variable on a project's development cycle. There is little notion in these models of the detailed practices that make up the totality. In mature organizations such practices are defined and can be analyzed individually and built up into a total estimate. Normally the errors in estimating these smaller components are smaller than the corresponding error at the total project level, and it is assumed that the summary effect of aggregating these smaller errors is still smaller than the error in the estimate at the total project level.

Reliability of Model Inputs

Even if a cost estimation model is statistically sound, the data on which it is based can have low validity. Often, managers do not have sufficient knowledge of crucial variables that must be entered into a model, such as the estimated size of various individual components of a system. In such instances, processes exist for increasing the accuracy of these data. For instance, Delphi techniques can be used by software engineers who have previous experience in developing various system components. The less experience an organization has with a particular component of a system, the less reliable is the size estimate for that component. Typically, component sizes are underestimated, with ruinous effects on the resources and schedule estimated for a project. Sometimes historical "fudge factors" are applied to account for underestimation, although a more rigorous data-based approach is recommended. To aid in identifying the potential risks in a software development project, it would also be beneficial to have reliable confidence bounds for different components of the estimated size or effort.

Statistical methods can be applied to develop prior probabilities (e.g., for Bayesian estimation models) from knowledgeable software engineers and to adjust these using historical data. These methods should be used not only to suggest the confidence that can be placed in an estimate, but also to indicate the components within a system that contribute most to inaccuracies in an estimate.

As projects progress during their life cycle from specifications of requirements to design to generation of code, the information on which estimates can be based grows more reliable: there is thus greater certainty in estimating from the architectural design of a system or the detailed design of each module than in estimating from textual statements. In short, the sources from which estimates can be developed change as the project continues through its development cycle. Each succeeding level of input is a more reliable indicator of the ultimate system size than are the inputs available in earlier stages of development. Thus the overall estimate of size, resources, and schedule potentially becomes more accurate in succeeding phases of a project. Yet it is important to determine the most accurate indicators of crucial parameters such as size, effort, and schedule very early in a project, when the least reliable data are available. As such, there is a need for statistically valid ways of developing model inputs from less reliable forms of data (these inputs must reliably estimate later measures that will be more valid inputs) and of estimating how much error is introduced into an estimate based on the reliability of the inputs.

Managing to Estimates

Complicating the ability to validate cost estimation models from historical data is the fact that project managers try to manage their projects to meet received estimates for cost, effort, schedule, and other such variables. Thus, an estimate affects the subsequent process, and historical data are made artificially more accurate by management decisions and other factors that are often masked in project data. For instance, projects whose required level of effort has been underestimated often survive on large amounts of unreported overtime put in by the development staff. Moreover, many managers are quite skilled at cutting functionality from a system in order to meet a delivery date. In the worst cases, engineers short-cut their ordinary engineering processes to meet an unrealistic schedule, usually with disastrous results. Techniques for modeling systems dynamics provide one way to characterize some of the interactions that occur between an estimate and the subsequent process that is generated by the estimate (Abdel-Hamid, 1991).

The validation of cost estimation models must be conducted with an understanding of such interactions between estimates and a project manager's decisions. Some of these dynamics may be usefully described by statistical models or by techniques developed in psychological decision theory (Kahneman et al., 1982). Thus, it may be possible to develop a statistical dynamic model (e.g., a multistage linear model) that characterizes the reliability of inputs to an estimate, the estimate itself, decisions made based on the estimate, the resulting performance of the project, measures that emerge later in the project, subsequent decision making based on these later measures, and the ultimate performance of the project. Such models would be valuable in helping project managers to understand the ramifications of decisions based on an initial estimate and also on subsequent periodic updates.

ASSESSMENT AND RELIABILITY

Reliability Growth Modeling

Many reliability models of varying degrees of plausibility are available to software engineers. These models are applied at either the testing stage or the field-monitoring stage. Most of the models take as input either failure time or failure count data and fit a stochastic process model to reflect reliability growth. The differences among the models lie principally in assumptions made based on the underlying stochastic process generating the data. A brief survey of some of the well-known models and their assumptions and efficacy is given in Abdel-Ghaly et al. (1986).

Although many software reliability growth models are described in the literature, the evidence suggests that they cannot be trusted to give accurate predictions in all cases and also that it is not possible to identify a priori which model (if any) will be trustworthy in a particular context. No doubt work will continue in refining these models and introducing "improved" ones. Although such work is of some interest, the panel does not believe that it merits extensive research by the statistical community, but thinks rather that statistical research could be directed more fruitfully to providing insight to the users of the models that currently exist.

The problem is validation of such models with respect to a particular data source, to allow users to decide which, if any, prediction scheme is producing accurate results for the actual software failure process under examination. Some work has been done on this problem (Abdel-Ghaly et al., 1986; Brocklehurst and Littlewood, 1992), using a combination of probability forecasting and sequential prediction, the so-called prequential approach developed by Dawid (1984), but this work has so far been rather informal. It would be helpful to have more procedures for assessing the accuracy of competing prediction systems that could then be used routinely by industrial software engineers without advanced statistical training.

Statistical inference in the area of reliability tends almost invariably to be of a classical frequentist kind, even though many of the models originate from a subjective Bayesian probability viewpoint. This unsatisfactory state of affairs arises from the sheer difficulty of performing the computations necessary for a proper Bayesian analysis. It seems likely that there would be profit in trying to overcome these problems, perhaps via the Gibbs sampling approach (see, e.g., Smith and Roberts, 1993).

Another fruitful avenue for research concerns the introduction of explanatory variables, so-called covariates, into software reliability growth models. Most existing models assume that no explanatory variables are available. This assumption is assuredly simplistic concerning testing for all but small systems involving short development and life cycles. For large systems (i.e., those with more than 100,000 lines of code) there are variables, other than time, that are very relevant. For example, it is typically assumed that the number of faults (found and unfound) in a system under test remains stable—i.e., that the code remains frozen—during testing. However, this is rarely the case for large systems, since aggressive delivery cycles force the final phases of development to overlap with the initial stages of system testing. Thus, the size of code and, consequently, the number of faults in a large system can vary widely during testing. If these changes in code size are not considered, the result, at best, is likely to be an increase in variability and a loss in predictive performance, and at worst, a poorly fitting model with unstable parameter estimates. Taking this logic one step further suggests the need to distinguish between new lines of code (new faults) and code coming from previous releases (old faults), and possibly the age of different parts of code. Of course, one can carry this logic to an extreme and have unwieldy models with many covariates. In practice, what is required is a compromise between the two extremes of having no covariates and having hundreds of them. This is where opportunities abound for applying state-of-the-art statistical modeling techniques. Described briefly below is a case study reported by Dalal and McIntosh (1994) dealing with reliability modeling when code is changing.

Example. Consider a new release of a large telecommunications system with approximately 7 million noncommentary source lines (NCSLs) and 400,000 lines of noncommentary new or changed source lines (NCNCSLs). For a faster delivery cycle, the source code used for system test was updated every night throughout the test period. At the end of each of 198 calendar days in the test cycle, the number of faults found, NCNCSLs, and the staff time spent on testing were collected. Figure 4 (top) portrays growth of the system as a function of staff time. The data are provided in Table 3.

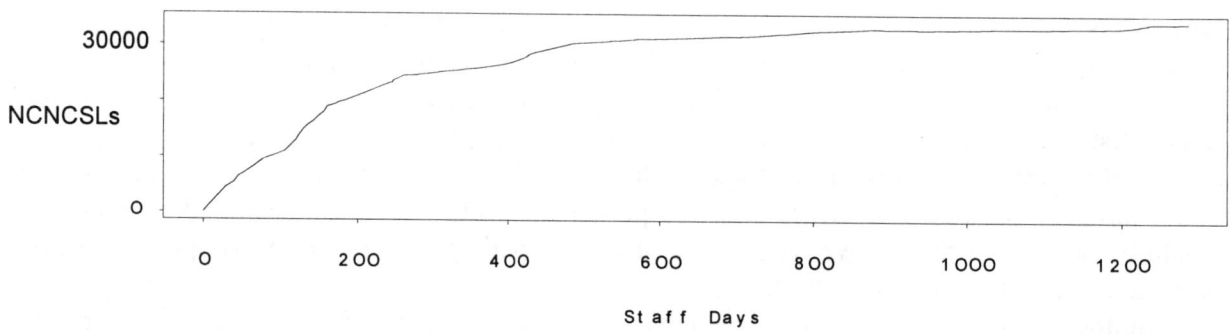

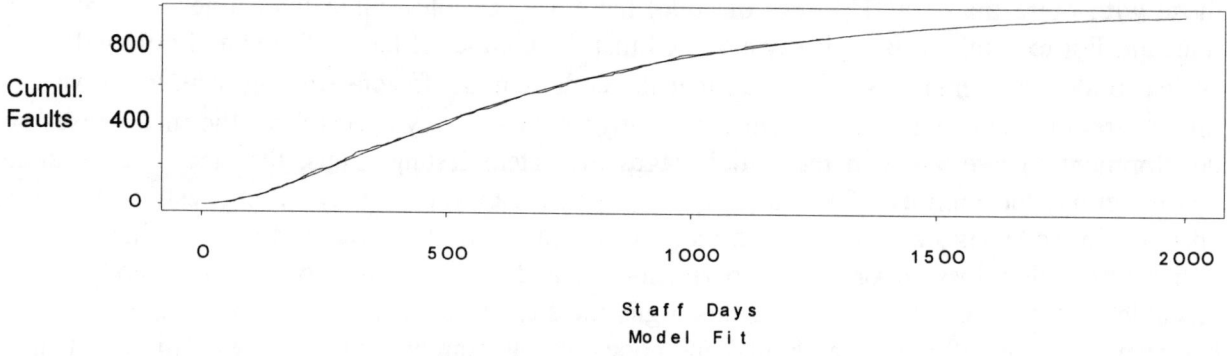

Figure 4. Plots of module size (NCNCSLs) versus staff time (days) for a large telecommunications software system (top). Observed and fitted cumulative faults versus staff time (bottom). The dotted line (barely visible) represents the fitted model, the solid line represents the observed data, and the dashed line (also difficult to see) is the extrapolation of the fitted model.

Table 3. Data on cumulative size (NCNCSLs), cumulative staff time (days), and cumulative faults for a large telecommunications system on 198 consecutive calendar days (with duplicate lines representing weekends or holidays).

Cum. Staff Days	Cum. Faults	Cum. NCNCSLs	Cum. Staff Days	Cum. Faults	Cum. NCNCSLs	Cum. Staff Days	Cum. Faults	Cum. NCNCSLs
0	0	0	334.8	231	261669	776.5	612	318476
4.8	0	16012	342.7	243	262889	793.5	621	320125
6	0	16012	350.5	252	263629	807.2	636	321774
6	0	16012	356.3	259	264367	811.8	639	321774
14.3	7	32027	360.6	271	265107	812.5	639	321774
22.8	7	48042	365.7	277	265845	829	648	323423
32.1	7	58854	365.7	277	265845	844.4	658	325072
41.4	7	69669	365.7	277	265845	860.5	666	326179
51.2	11	80483	374.9	282	266585	876.7	674	327286
51.2	11	80483	386.5	290	267325	892	679	328393
51.2	11	80483	396.5	300	268607	895.5	686	328393
60.6	12	91295	408	310	269891	895.5	686	328393
70	13	102110	417.3	312	271175	910.8	690	329500
79.9	15	112925	417.3	312	271175	925.1	701	330608
91.3	20	120367	417.3	312	271175	938.3	710	330435
97	21	127812	424.9	321	272457	952	720	330263
97	21	127812	434.2	326	273741	965	729	330091
97	21	127812	442.7	339	275025	967.7	729	330091
97	21	127812	451.4	346	276556	968.6	731	330091
107.7	22	135257	456.1	347	278087	981.3	740	329919
119.1	28	142702	456.1	347	278087	997	749	329747
127.6	40	150147	456.1	347	278087	1013.9	759	330036
135.1	44	152806	460.8	351	279618	1030.1	776	330326
135.1	44	152806	466	356	281149	1044	781	330616
135.1	44	152806	472.3	359	283592	1047	782	330616
142.8	46	155464	476.4	362	286036	1047	782	330616
148.9	48	158123	480.9	367	288480	1059.7	783	330906
156.6	52	160781	480.9	367	288480	1072.6	787	331196
163.9	52	167704	480.9	367	288480	1085.7	793	331486
169.7	59	174626	486.8	374	290923	1098.4	796	331577
170.1	59	174626	495.8	376	293367	1112.4	797	331669
170.6	59	174626	505.7	380	295811	1113.5	798	331669
174.7	63	181548	516	392	298254	1114.1	798	331669
179.6	68	188473	526.2	399	300698	1128	802	331760
185.5	71	194626	527.3	401	300698	1139.1	805	331852
194	88	200782	527.3	401	300698	1151.4	811	331944
200.3	93	206937	535.8	405	303142	1163.2	823	332167
200.3	93	206937	546.3	415	304063	1174.3	827	332391
200.3	93	206937	556.1	425	305009	1174.3	827	332391
207.2	97	213093	568.1	440	305956	1174.3	827	332391
211.9	98	219248	577.2	457	306902	1184.6	832	332615
217	105	221355	578.3	457	306902	1198.3	834	332839
223.5	113	223462	578.3	457	306902	1210.3	836	333053
227	113	225568	587.2	467	307849	1221.1	839	333267
227	113	225568	595.5	473	308795	1230.5	842	333481
227	113	225568	605.6	480	309742	1231.6	842	333481
234.1	122	227675	613.9	491	310688	1231.6	842	333481
241.6	129	229784	621.6	496	311635	1240.9	844	333695
250.7	141	233557	621.6	496	311635	1249.5	845	333909
259.8	155	237330	621.6	496	311635	1262.2	849	335920
268.3	166	241103	623.4	496	311635	1271.3	851	337932
268.3	166	241103	636.3	502	311750	1279.8	854	339943
268.3	166	241103	649.7	517	311866	1281	854	339943
277.2	178	244879	663.9	527	312467	1281	854	339943
285.5	186	247946	675.1	540	313069	1287.4	855	341955
294.2	190	251016	677.4	543	313069	1295.1	859	341967
295.7	190	251016	677.9	544	313069	1304.8	860	341979
298	190	254086	688.4	553	313671	1305.8	865	342073
298	190	254086	698.1	561	314273	1313.3	867	342168
298	190	254086	710.5	573	314783	1314.4	867	342168
305.2	195	257155	720.9	581	315294	1314.4	867	342168
312.3	201	260225	731.6	584	315805	1320	867	342262
318.2	209	260705	732.7	585	315805	1325.3	867	342357
328.9	224	261188	733.6	585	315805	1330.6	870	342357
334.8	231	261669	746.7	586	316316	1334.2	870	342358
334.8	231	261669	761	598	316827	1336.7	870	342358

SOURCE: Dalal and McIntosh (1994).

Assume that the testing process is observed at time $t_i, i = 0,...,h$, and at any given time, the amount of time it takes to find a specific "bug" is exponential with rate m. At time t_i, the total number of faults remaining in the system is Poisson with mean l_{i+1}, and NCNCSL is increased by an amount C_i. This change adds a Poisson number of faults with mean proportional to C, say qC_i. These assumptions lead to the mass balance equation, namely, that the expected number of faults in the system at t_i (after possible modification) is the expected number of faults in the system at t_{i-1} adjusted by the expected number found in the interval (t_{i-1}, t_i) plus the faults introduced by the changes made at t_i:

$$l_{i+1} = l_i e^{-m(t_i - t_{i-1})} + qC_i,$$

for $i = 1,...h$. Note that q represents the number of new faults entering the system per additional NCNCSL, and l_1 represents the number of faults in the code at the start of system test. Both of these parameters make it possible to differentiate between the new code added in the current release and the older code. For the data at hand, the estimated parameters are $q = 0.025$, $m = 0.002$, and $l_1 = 41$. The fitted and the observed data are plotted against staff time in Figure 4 (bottom). The fit is evidently very good. Of course assessing the model on independent or new data is required for proper validation.

The efficacy of creating a statistical model is now examined. The estimate of q is highly significant, both statistically and practically, showing the need for incorporating changes in NCNCSLs as a covariate. Its numerical value implies that for every additional 10,000 NCNCSLs added to the system, 25 faults are being added as well. For these data, the predicted number of faults at the end of the test period is Poisson distributed with mean 145. Dividing this quantity by the total NCNCSLs gives 4.2 per 10,000 NCNCSLs as an estimated field fault density. These estimates of the incoming and outgoing quality are very valuable in judging the efficacy of system testing and for deciding where resources should be allocated to improve the quality. Here, for example, system testing was effective in that it removed 21 of every 25 faults. However, it raises another issue: 25 faults per 10,000 NCNCSLs entering system test may be too high and a plan ought to be considered to improve the incoming quality. ∎

None of the above conclusions could have been made without using a statistical model. These conclusions are valuable for controlling and improving the reliability testing process. Further, for this analysis it was essential to have a covariate other than time.

Influence of the Development Process on Software Dependability

As noted above, surprisingly little use has been made of explanatory variable models, such as proportional hazards regression, in the modeling of software dependability. A major reason, the panel believes, is the difficulty that software engineers have in identifying variables that can play a genuinely explanatory role. Another difficulty is the comparative paucity of data owing to the difficulties of replication. Thus, for example, for purposes of identifying those attributes of the software development process that are drivers of the final product's dependability, it is very difficult to obtain something akin to a "random sample" of "similar" subject programs. Those issues are not unlike the ones faced in other contexts where these techniques are used, for example, in medical trials, but they seem particularly acute for evaluation of software dependability.

A further problem is that the observable in this software development application is a realization of a stochastic process, and not merely of a lifetime random variable. Thus there seems to be an opportunity for research into models that, on the one hand, capture current understanding of the nature of the growth in reliability that takes place as a result of debugging and, on the other hand, allow input about the nature of the development process or the architecture of the product.

Influence of the Operational Environment on Software Dependability

It can be misleading to talk of the reliability of a program: as is the case for the reliability of hardware, the reliability of a program depends on the nature of its use. For software, however, one does not have the simple notions of stress that are sometimes plausible in the hardware context. It is thus not possible to infer the reliability of a program in one environment from evidence of the program's failure behavior in another. This is a serious difficulty for several reasons.

First, one would like to be able to predict the operational reliability of a program from test data. The simplest approach at present is to ensure that the test environment, that is, the type of usage, is exactly similar to, or differs in known proportions for specified strata from, the operational environment. Real software testing regimes are often deliberately made to be different from operational ones, since it is claimed that in this way reliability can be achieved more efficiently: this argument is similar to that for hardware stress testing but is much less convincing in the software context.

A further reason to be interested in this problem of inferring program reliability is that most software gets broadly distributed to diverse locations and is used very differently by different users: there is great disparity in the population of user environments. Vendors would like to be able to predict different users' perceptions of a product's reliability, but it is clearly impractical to replicate in a test every different possible operational environment. Vendors would also like to be able to predict the characteristics of a *population* of users. Thus it might be expected that a less disparate population of users would be preferable to a more disparate one: in the former case, for example, problems reported at different sites might be similar and thus be less expensive to fix.

Explanatory variable modeling may play a useful role if suitably informative, measurable attributes of operational usage can be identified. There may be other ways of forming stochastic characterizations of operational environments. Markov models of the successive activation of modules, or of functions, have been proposed (Littlewood, 1979; Siegrist, 1988a,b) but have not been widely used. Further work on such approaches, and on the problems of statistical inference associated with them, could be promising.

Safety-Critical Software and the Problem of Assuring Ultrahigh Dependability

It seems clear that computers will play increasingly critical roles in systems upon which human lives depend. Already, systems are being built that require extremely high dependability—a figure of 10^{-9} probability of failure per hour of flight has been stated as the requirement for recent fly-by-wire systems in civil aircraft. There are clear limitations to the levels of dependability that can be achieved when we are building systems of a complexity that precludes claims that they are free of design faults. More importantly, even if we were able to build a system to meet a requirement for ultrahigh dependability, we could have only low confidence that we had achieved that goal, because the problem of assessing these levels is such that it would be impractical to acquire sufficient supporting evidence (Littlewood and Strigini, 1993).

Although a complete solution to the problem of assessing ultrahigh dependability is not anticipated, there is certainly room for improving on what can be done currently. Probabilistic and statistical problems abound in this area, and it is necessary to squeeze as much as possible from relatively small amounts of often disparate evidence. The following are some of the areas that could benefit from investigation.

Design Diversity, Fault Tolerance, and General Issues of Dependence

One promising approach to the problem of achieving high dependability (here reliability and/or safety) is design diversity: building two or more versions of the required program and allowing an adjudication mechanism (e.g., a voter) to operate at run-time. Although such systems have been built and are in operation in safety-critical contexts, there is little theoretical understanding of their behavior in operation. In particular, the reliability and safety models are quite poor.

For example, there is ample evidence (Knight and Leveson, 1986) that, in the presence of design faults, one cannot simply assume that different versions will fail independently of one another. Thus the simple hardware reliability models that involve mere redundancy, and assume independence of component failures, cannot be used. It is only quite recently that probability modeling has started to address this problem seriously (Eckhardt and Lee, 1985; Littlewood and Miller, 1989). These models provide a formal conceptual framework within which it is possible to reason about the subtle issues of conditional independence involved in the failure processes of design-diverse systems. However, they provide little quantitative practical assistance to a software designer or evaluator.

Further probabilistic modeling is needed to elucidate some of the complex issues. For example, little attention has been paid to modeling the full fault tolerant system, involving diversity and adjudication. In particular, the properties of the stochastic process of failures of such systems are not understood. If, as seems likely, individual versions of a program in a real-time control system exhibit clusters of failures in time, how does the cluster process of the system relate to the cluster processes of the individual versions? Although such issues seem narrowly technical, they are vitally important in the design of real systems, whose physical integrity may be sufficient to survive one or two failed input cycles, but not many.

Another area that has had little work is probabilistic modeling of different possible adjudication mechanisms and their failure processes.

Judgment and Decision-making Framework

Although probability seems to be the most appropriate mechanism for representing uncertainty about system dependability, other candidates such as Shafer-Dempster and possibility theories might be plausible alternatives in safety-critical contexts where quantitative measures are required in the absence of data—for example, when one is forced to rely on the engineering judgment of an expert. Further work is needed to elucidate the relative advantages and disadvantages of the different approaches applicable in the software engineering domain.

There is evidence that human judgment, even in "hard" sciences such as physics, can be seriously in error (Henrion and Fischhoff, 1986): people seem to make consistent errors and tend to be optimistic in their own judgment regarding their likely error. It is likely that software engineering judgments are similarly fallible, and so this area calls for some statistical experimentation. In addition, it would be beneficial to have formal mechanisms for assessing whether judgments are well calibrated and for recalibrating judgment and prediction schemes (of humans or models) that have been shown to be inaccurate. This problem has some similarity to the problems of validating software reliability models, already mentioned, in which prequential likelihood plays a vital role. It also bears on more general applications of Bayesian modeling where elicitation of a priori probability values is required.

It seems inevitable that reasoning and judgment about the fitness of safety-critical systems will depend on evidence that is disparate in nature. Such evidence could include data on failures, as in reliability growth models; human expert judgment; results regarding the efficacy of development processes; information about the architecture of a system; or evidence from formal verification. If the required judgment depends on a numerical assessment of a system's dependability, there are clearly important issues concerning the composition of very different kinds of evidence from different sources. These issues may, indeed, be overriding when it comes to choosing among the different ways of representing uncertainty. The Bayes theorem, for example, may provide an easier way than does possibility theory to combine information from different sources of uncertainty.

A particularly important problem concerns the way in which deterministic reasoning can be incorporated into the final assessment of a system. Formal methods of achieving dependability are becoming increasingly important. Such methods range from formal notations, which assist in the elicitation and expression of requirements, to full mathematical verification of the correspondence between a formal specification and an implementation. One view is that these approaches incorporating deterministic reasoning to system development remove a particular type of uncertainty, leaving others untouched (uncertainty about the completeness of a formal specification, the possibility of incorrect proof, and so on). One should factor into the final assessment of a system's dependability the contribution from such deterministic, logical evidence, nevertheless keeping in mind that there is an irreducible uncertainty in one's possible knowledge of the failure behavior of a system.

Structural Modeling Issues

Concerns about the safety and reliability of software-based systems necessarily arise from their inherent complexity and novelty. Systems now being built are so complex that they cannot be guaranteed to be free from design faults. The extent to which confidence can be carried over from the building of previous systems is much more limited in software engineering than in "real" engineering, because software-based systems tend to be characterized by a great deal of novelty.

Designers need help in making decisions throughout the design process, especially at the very highest level. Real systems are often difficult to assess because of early decisions regarding how much system control will depend on computers, hardware, and humans. For the Airbus A320, for example, the early decision to place a high level of trust in the computerized fly-by-wire system meant that this system (and thus its software) needed to have a better than 10^{-9} probability of failure in a typical flight. Stochastic modeling might aid in such high-level design decisions so that designers can make "what if" calculations at an early stage.

Experimentation, Data Collection, and General Statistical Techniques

A dearth of data has been a problem in much of safety-critical software engineering since its inception. Only a handful of published data sets exists even for the software reliability growth problem, which is by far the most extensively developed aspect of software dependability assessment. When the lack of data arises from the need for confidentiality—industrial companies are often reluctant to allow access to data on software failures because of the possibility that people may think less highly of their products—little can be done beyond making efforts to resolve confidentiality problems. However, in some cases the available data are sparse because there is no statistical expertise on hand to advise on ways in which data can be collected cost-effectively. It may be worthwhile to attempt to produce general guidelines for data collection that address the specific difficulties of the software engineering problem domain.

With notable exceptions (Eckhardt et al., 1991; Knight and Leveson, 1986), experimentation has so far played a low-key role in software engineering research. Somewhat surprisingly, in view of its difficulty and cost, the most extensive experimentation has investigated the efficacy of design diversity. Other areas where experimental approaches seem feasible and should be encouraged include the obvious and general question of which software development methods are most cost-effective in producing software products with desirable attributes such as dependability. Statistical advice on the design of such experiments would be essential; it might also be the case that innovation in the design of experiments could make feasible some investigations that currently seem too expensive to contemplate: the main problem arises from the need for replication over many software products.

On the other hand, areas where experiments can be conducted without the replication problem being overwhelming involve the investigation of quite restricted hypotheses about the effectiveness of specific techniques. For example, experimentation could address whether the techniques that are claimed to be effective for achieving reliability (i.e., effectiveness of debugging) are significantly better than those, such as operational testing, that will allow reliability to be measured.

SOFTWARE MEASUREMENT AND METRICS

Measurement is at the foundation of science and engineering. An important goal shared by software engineers and statisticians is to derive reliable, reproducible, and accurate measures of software products and processes. Measurements are important for assessing the effects of proposed "improvements" in software production, whether they be technological or process oriented. Measurements serve an equally important role in scheduling, planning, resource allocation, and cost estimation (see the first section in this chapter).

Early pioneering work by McCabe (1976) and Halstead (1977) seeded the field of software metrics; an overview is provided by Zuse (1991). Much of the attention in this area has focused on static measurements of code. Less attention has been paid to dynamic measurements of software (e.g., measuring the connectivity of software modules under operating conditions) and aspects of the software production process such as software reuse, especially in systems employing object-oriented languages.

The most widely used code metric, the NCSL (noncommentary source line), is often used as a surrogate for functionality. Surprisingly, since software is now nearly 50 years old, standards for counting NCSLs remain elusive in practice. For example, should a single, two-line statement in C language count as one NCSL or two?

Counts of tokens (operators or operands), delimiters, and branching statements are used as other static metrics. Although some of these are clearly measures of software size, others purport to measure more subtle notions of software complexity and structure. It has been observed that all such metrics are highly correlated with size. At the panel's information-gathering forum, Munson (1993) concluded that current software metrics capture approximately three "independent" features of a software module: program control, program size, and data structure. A statistical (principal-components) analysis of 13 metrics on HAL programs in the space shuttle program was the key to this finding. While one might argue that performing a common statistical decomposition of multivariate data is hardly novel, it most certainly is in software engineering. The important implication of that finding is that there are features of software that are not being captured by the existing battery of software metrics (e.g., cohesion and coupling)—and if these are key differentiators of potentially high- and low-fault programs, there is no way that an analysis of the available metrics will highlight this condition. On the other side of the ledger, the statistical costs of including "noisy" versions of the same (latent) variable in models and analysis methods that are based on these metrics, such as cost estimation, seem not to have been appreciated. Subset selection methods (e.g., Mallows, 1973) provide one way to assess variable redundancy and the effect on fitted models, but other approaches that use judgment composites, or composites based on other bodies of data (Tukey, 1991), will often be more effective than discarding metrics.

Metrics typically involve processes or products, are subjective or objective, and involve different types of measurement scales, for example, nominal, ordinal, interval, or ratio. An objective metric is a measurement taken on a product or process, usually on an interval or ratio scale. Some examples include the number of lines of code, development time, number of software faults, or number of changes. A subjective metric may involve a classification or qualification based on experience. Examples include the quality of use of a method or the experience of the programmers in the application or process.

One standard for software measurement is the Basili and Weiss (1984) Goal/Question/Metric paradigm, which has five parameters:

1. An object of the study—a process, product, or any other experience model;
2. A focus—what information is of interest;
3. A point of view—the perspective of the person needing the information;
4. A purpose—how the information will be used; and
5. A determination of what measurements will provide the information that is needed.

The results are studied relative to a particular environment.

5
Statistical Challenges

In comparison with other engineering disciplines, software engineering is still in the definition stage. Characteristics of established disciplines include having defined, time-tested, credible methodologies for disciplinary practice, assessment, and predictability. Software engineering combines application domain knowledge, computer science, statistics, behavioral science, and human factors issues. Statistical research and education challenges in software engineering involve the following:

- Generalizing particular experimental results to other settings and projects,
- Scaling up results obtained in academic studies to industrial settings,
- Combining information across software engineering projects and studies,
- Adopting exploratory data analysis and visualization techniques,
- Educating the software engineering community as to statistical approaches and data issues,
- Developing analysis methods to cope with qualitative variables,
- Providing models with the appropriate error distributions for software engineering applications, and
- Improving accelerated life testing.

The following sections elaborate on certain of these challenges.

SOFTWARE ENGINEERING EXPERIMENTAL ISSUES

Software engineering is an evolutionary and experimental discipline. As argued forcefully by Basili (1993), it is a laboratory or experimental science. The term "experimental science" has different meanings for engineers and statisticians. For engineers, software is experimental because systems are built, studied, and evaluated based on theory. Each system investigates new ideas and advances the state of the art. For statisticians, the purpose of experiments is to gather statistically valid evidence about the effects of some factor, perhaps involving the process, methodology, or code in a system.

There are three classes of experiments in software engineering:

- Case studies,
- Academic experiments, and
- Industrial experiments.

Case studies are perhaps the most common and involve an "experiment" on a single large-scale project. Academic experiments usually involve a small-scale experiment, often on a program or methodology, typically using students as the experimental subjects. Industrial experiments fall somewhere between case studies and academic experiments. Because of the expense and difficulty of performing extensive controlled experiments on software, case studies are often resorted to. The ideal situation is to be able to take advantage of real-world industrial operations while having as much control as is feasible. Much of the present work in this area is at best anecdotal and would benefit greatly from more rigorous statistical advice and control. The panel foresees an opportunity for innovative work on combining information (see below) from relatively disparate experiences.

Conducting statistically valid software experiments is challenging for several reasons:

- The software production process is often chaotic and uncontrolled (i.e., immature);
- Human variability is a complicating factor; and
- Industrial experiments are very costly and therefore must produce something useful.

Many variables in the software production process are not well understood and are difficult to control for. For software engineering experiments, the factors of interest include the following:

- "People" factors: number, level, organization, process experience;
- Problem factors: application domain, constraints, susceptibility to change;
- Process factors: life cycle model, methods, tools, programming language;
- Product factors: deliverables, system size, system reliability, portability; and
- Resource factors: target and development machines, calendar time, budget, existing software, and so on.

Each of these characteristics must be modeled or controls done for the experiment to be valid.

Human variability is particularly challenging, given that the difference in quality and productivity between the best and worst programmers may be 20 to 1. For example, in an experiment comparing batch versus interactive computing, Sackman (1970) observed differences in ability of up to 28 to 1 in programmers performing the same task. This variation can overwhelm the effects of a change in methodology that may account for a 10% to 15% difference in quality or productivity.

The human factor is so strongly integrated with every aspect of the subjective discipline of software engineering that it alone is the prime driver of issues to be addressed. The human factor creates issues in the process, the product, and the user environment. Measurements of the objects (the product and the process) are obscured when qualified by the attributes (ambiguous requirements and productivity issues are key examples). Recognizing and characterizing the human attributes within the context of the software process are key to understanding how to include them in system and statistical models.

The capabilities of individuals strongly influence the metrics collected throughout the software production process. Capabilities include experience, intelligence, familiarity with the application domain, ability to communicate with others, ability to envision the problem spatially, and ability to verbally describe that spatial understanding. Although not scientifically founded, anecdotal information supports the incidence of these capabilities (Curtis, 1988).

For software engineering experiments, the key problems involve small sample sizes, high variability, many uncontrolled factors, and extreme difficulty in collecting experimental data. Traditional statistical experimental designs, originally developed for agricultural experiments, are not well suited for software engineering. At the panel's forum, Zweben (1993) discussed an interesting example of an experiment from object-oriented programming, involving a fairly complex design and analysis. Object-oriented programming is an approach that is sweeping the software industry, but for which much of the supporting evidence is anecdotal.

Example. The purpose of the software design and analysis experiment was to gather statistically valid evidence about the effect—on effort and quality—of using the principles of abstraction, encapsulation, and layering to enhance components of software systems. The experiment was divided into two types of tasks:

1. Enhancing an existing component to provide additional functionality, and
2. Modifying a component to provide different functionality.

The experimental subjects were students in graduate classes on software component design and development. The two approaches for this maintenance problem are "white box," which involves modifying the old code to get the new functionality, and "black box," which involves layering on the new functionality. The experiments were designed to detect, for each task, differences between the two approaches in the time required to make the modification and in the number of associated faults uncovered. Three experiments were conducted. Experiment A involved an unbounded queue component. The subjects were given a basic Ada package implementing *enque, deque,* and *is empty,* and the task was to implement the operators *add, copy, clear, append,* and *reverse*. The subject was instructed to keep track of the time spent in designing, coding, testing, and debugging each operator, and also the associated number of bugs uncovered in each task. The tasks were completed in two ways: by directly implementing new operations using the representation of the queue, and by layering on the new operators as capabilities. Experiment B involved a partial map component, and experiment C involved an almost constant map component. Given that in experiments involving students, the results may be invalidated by problems with data integrity, for this experiment the student participants were told that the results of the experiment would have no effect on course grades. The code was validated by an instructor to ensure that there were no lingering defects. The experimental plan was conducted using a crossover design. Each subject implemented the enhancements twice, using both the white box and the black box methods. This particular experimental design could test for the treatment (layering or not) effect and treatment by sequence interaction. The subject differences were nested within the sequences, and the sequences were counterbalanced based on experience level. The carryover effect of the first treatment influences the choice regarding the correct way of testing for treatment effects.

The statistical model used to represent the behavior in the number of bugs was sophisticated as well, an overdispersed log linear model. The use of this model allowed for an analysis of nonnormal response data while also preventing invalid inferences that would have occurred had overdispersion not been taken into account. Indeed, only experiment B displayed a significant treatment effect after adjustment for overdispersion. ∎

COMBINING INFORMATION

The results of many diverse software projects and studies tend to lead to more confusion than insight. The software engineering community would benefit if more value were gained from the work that is being done. To the extent that projects and studies focus on the same end point, statistics can help to fuse the independent results into a consistent and analytically justifiable story.

The statistical methodology that addresses the topic of how to fuse such independent results is relatively new and is termed "combining information"; a related set of tools is provided by meta-analysis. An excellent overview of this methodology was produced by a CATS panel and documented in an NRC report (NRC, 1992) that is now available as an American Statistical Association publication (ASA, 1993). The report documents various approaches to the problem of how to combine information and describes numerous specific applications. One of the recommendations made in it (p. 182) is crucial to achieving advances in software engineering:

> The panel urges that authors and journal editors attempt to raise the level of quantitative explicitness in the reporting of research findings, by publishing summaries of appropriate quantitative measures on which the research conclusions are based (e.g., at a minimum: sample sizes, means, and standard deviations for all variables, and relevant correlation matrices).

It is not sensible to merely combine p-values from independent studies. It is clearly better to take weighted averages of effects when the weights account for differences in size and sensitivity across the studies to be combined.

Example. Kitchenham (1991) discusses an issue in cost estimation that involves looking across 10 different sources consisting of 17 different software projects. The issue is whether the exponent β in the basic cost estimation model, $\mathit{effort} \propto \mathit{size}^\beta$, is significantly different from 1. The usual interpretation of β is the "overhead introduced by product size," so that a value greater than 1 implies that relatively more effort is required to produce large software systems than to produce smaller ones. Many cite such "diseconomies of scale" in software production as evidence in support of their models and tools.

The 17 software projects are listed in Table 4. Fortunately, the cited sources contain both point estimates (b) of the exponent and its estimated standard error. These summary statistics can be used to estimate a common exponent and ultimately test the hypothesis that it is different from 1.

Table 4. Reported and derived data on 17 projects concerned with cost estimation.

Study	b	SE (b)	Var (b)	w
Bai-Bas	0.951	0.068	0.004624	21.240
Bel-Leh	1.062	0.101	0.010200	18.990
Your	0.716	0.230	0.052900	10.490
Wing	1.059	0.294	0.086440	7.758
Kemr	0.856	0.177	0.031330	13.550
Boehm.Org	0.833	0.184	0.033860	13.100
Boehm.semi	0.976	0.133	0.017690	16.630
Boehm.Emb	1.070	0.104	0.010820	18.770
Kit-Tay.ICL	0.472	0.323	0.104300	6.813
Kit-Tay.BTSX	1.202	0.300	0.090000	7.550
Kit-Tay.BTSW	0.495	0.185	0.034220	13.040
DS1.1	1.049	0.125	0.015630	17.220
DS1.2	1.078	0.105	0.011020	18.700
DS1.3	1.086	0.289	0.083520	7.938
DS2.New	0.178	0.134	0.017960	16.550
DS2.Ext	1.025	0.158	0.024960	14.830
DS3	1.141	0.077	0.005929	20.670

SOURCE: Reprinted, with permission, from Kitchenham (1992). © 1992 by National Computing Centre, Ltd.

Following the NRC recommendations on combining information across studies (NRC, 1992), the appropriate model (the so-called random effects model in meta-analysis) allows for a systematic difference between projects (e.g., bias in data reporting, management style, and so on) that averages to zero. Under this model, the overall exponent is estimated as a weighted average of the individual exponents where the weights have the form $w_i = \text{var}(b_i) + \tau^2$ and the common between-project component of variance is estimated by

$$t^2 = \max\left\{0, \frac{Q-(k-1)}{\Sigma w_i - \left(\Sigma w_i^2 / \Sigma w_i\right)}\right\},$$

where $Q = \Sigma w_i (b_i - \hat{b})^2$. The statistic Q is itself a test of the homogeneity of projects and under a normality assumption is distributed as χ^2_{k-1}. For these data one obtains $Q = 55.19$, which strongly indicates heterogeneity across projects. Although the random effects model anticipates such heterogeneity, other approaches that model the differences between projects (e.g., regression models) may be more informative. Since no explanatory variables are available, this discussion proceeds using the simpler model.

The estimated between-project component of variance is $t^2 = 0.0425$, which is surprisingly large and is perhaps highly influenced by two projects with b's less than 0.5. Combining this estimate with the individual within-project variances leads to the weights given in the final column of Table 4. Thus the overall estimated exponent is $\hat{b} = 0.911$ with estimated standard error $s = 0.0640$ ($= \sqrt{1/\Sigma w_i}$). Combining these two estimates leads readily to a 95% confidence interval for β of (0.78, 1.04). Thus the data in these studies do not support the diseconomies-of-scale argument. ∎

Even better than published summaries would be a central repository of the data arising from a study. This information would allow assessment of various determinations of similarities between studies, as well as potential biases. The panel is aware of several initiatives to build such data repositories. The proposed National Software Council has as one of its primary responsibilities the construction and maintenance of a national software measurements database. At the panel's forum, a specialized database on software projects in the aeronautics industry was also discussed (Keller, 1993).

An issue related to combining information from diverse sources concerns the translation to industry of small experimental studies and/or published case studies done in an academic environment. Serious doubts exist in industry as to the upward scalability of most of these studies because populations, project sizes, and environments are all different. Expectations differ regarding quality, and it is unclear whether variables measured in a small study are the variables in which industry has an interest. The statistical community should develop stochastic models to propagate uncertainty (including variability assessment) on different control factors so that adjustments and predictions applicable to industry-level environments can be made.

VISUALIZATION IN SOFTWARE ENGINEERING

Scientific visualization is an emerging technology that is driven by ever-decreasing hardware prices and the associated increasing sophistication of visualization software. Visualization involves the interactive pictorial display of data using graphics, animation, and sound. Much of the recent progress in visualization has come from the application of computer graphics to three-dimensional image analysis and rendering. Data visualization, a subset of scientific visualization, focuses on the display and analysis of abstract data. Some of the earliest and best-known examples of data visualization involve statistical data displays.

The motivation for applying visualization to software engineering is to understand the complexity, multidimensionality, and structure embodied in software systems. Much of the original research in software visualization—the use of typography, graphic design, animation, and cinematography to facilitate the understanding and enhancement of software systems—was performed by computer scientists interested in understanding algorithms, particularly in the context of education. Applying the quantitative focus of statistical graphics methods to currently popular scientific visualization techniques is a fertile area for research.

Visualizing software engineering data is challenging because of the diversity of data sets associated with software projects. For data sets involving software faults, times to failure, cost and effort predictions, and so on, there is a clear statistical relationship of interest. Software fault density may be related to code complexity and to other software metrics. Traditional techniques for visualizing statistical data are designed to extract quantitative relationships between variables. Other software engineering data sets such as the execution trace of a program (the sequence of statements executed during a test run) or the change history of a file are not easily visualized using conventional data visualization techniques. The need for relevant techniques has led to the development of specialized domain-specific visualization capabilities peculiar to software systems. Applications include the following:

- Configuration management data (Eick et al., 1992b),
- Function call graphs (Ganser et al., 1993),
- Code coverage,
- Code metrics,
- Algorithm animation (Brown and Hershberger, 1992; Stasko, 1993),
- Sophisticated typesetting of computer programs (Baecker and Marcus, 1988),
- Software development process,
- Software metrics (Ebert, 1992), and
- Software reliability models and data.

Some of these applications are discussed below.

Configuration Management Data

A rich software database suitable for visualization involves the code itself. In production systems, the source code is stored in configuration management databases. These databases contain a complete history of the code with every source code change recorded as a modification request. Along with the affected lines, the source code database usually contains other information such as the identity of the programmer making the changes, date the changes were submitted, reason for the change, and whether the change was meant to add functionality or fix a bug. The variables associated with source code may be continuous, categorical, or binary. For a line in a computer program, when it was written is (essentially) continuous, who wrote it is categorical, and whether or not the line was executed during a regression test is binary.

Example. Figure 1 (see "Implementation" in Chapter 3) shows production code written in C language from a module in AT&T's 5ESS switch (Eick, 1994). In the display, row color is tied to the code's age: the most recently added lines are in red and the oldest in blue, with a color spectrum in between. Dynamic graphics techniques are employed for increasing the effectiveness of the display. There are five interactive views of data in Figure 1:

1. The rows corresponding to the text lines,
2. The values on the color scale,
3. The file names above the columns,
4. The browser windows, and
5. The bar chart beneath the color scale.

Each of the views is linked, united through the use of color, and activated by using a mouse pointer. This mode of manipulating the display, called brushing by Becker and Cleveland (1987) and by Becker et al. (1987), is particularly effective for exploring software development data. ∎

Function Call Graphs

Perhaps the most common visualization of software is a function call graph as shown in Figure 5. Function call graphs are a widely used, visual, tree-like display of the function calls in a piece of code. They show calling relationships between modules in a system and are one representation of software structure. A problem with function call graphs is that they become overloaded with too much information for all but the smallest systems. One approach to improving the usefulness of function call graphs might involve the use of dynamic graphics techniques to focus the display on the visually informative regions.

Test Code Coverage

Another interesting example of source code visualization involves showing test suite code coverage. Figure 6 shows the statement coverage and execution "hot spots" for a program that has been run through its regression test. The row indentation and line length have been turned off so that each line receives the same amount of visual space. The most frequently executed lines are shown in red and the least frequently in blue, with a color spectrum in between. There are two special colors: the black lines correspond to nonexecutable lines of C code such as comments, variable declarations, and functions, and the gray lines correspond to the executable lines of code that were not executed. These are the lines that the regression test missed.

Code Metrics

As discussed in Chapter 4 (in the section "Software Measurement and Metrics"), static code metrics attempt to quantify and measure the complexity of code. These metrics are used to identify portions of programs that are particularly difficult and are likely to be subject to defects. One visualization method for displaying code complexity metrics uses a space-filling representation (Baker and Eick, 1995). Taking advantage of the hierarchical structure of code, each subsystem, module, and file is tiled on the display, which shows them as nested, space-filling rectangles with area, color, and fill encoding software metrics. This technique can display the relative sizes of a system's components, the relative stability of the components, the location of new functionality, the location of error-prone code with many fixes to identified faults, and, using animation, the historical evolution of the code.

Example. Figure 7 displays the AT&T 5ESS switching code using the SeeSys™ system, a dynamic graphics metrics visualization system. Interactive controls enable the user to manipulate the display, reset the colors, and zoom in on particular modules and files, providing an interactive software data analysis environment. The space-filling representation:

- Shows modules, files, and subsystems in context;
- Provides an overview of a complete software system; and
- Applies statistical dynamic graphics techniques to the problem of visualizing metrics. ■

A major difference in the use of graphics in scientific visualization and statistics is that for the former, graphs are the end, whereas for the latter, they are more often the means to an end. Thus visualizations of software are crucial to statistical software engineering to the extent that they facilitate description and modeling of software engineering data. Discussed below are some possibilities related to the examples described in this chapter.

The rainbow files in Figure 1 suggest that certain code is changed frequently. Frequently changed code is often error-prone, difficult to maintain, and problematic. Software engineers often claim that code, or people's understanding of it, decays with age. Eventually the code becomes unmaintainable and must be rewritten (re-engineered). Statistical models are needed to characterize the normal rate of change and therefore determine whether the current files are unusual. Such models need to take account of the number of changes, locations of faults, type of functionality, past development patterns, and future trends. For example, a common software design involves having a simple main routine that calls on several other procedures to invoke needed functionality. The main routine may be changed frequently as programmers modify small snippets of code to access large chunks of new code that is put into other files. For this code, many simple, small changes are normal and do not indicate maintenance problems. If models existed, then it would be possible to make quantitative comparisons between files rather than the qualitative comparisons that are currently made.

Figure 5 suggests some natural covariates and models for improving the efficiency of software testing. Current compiler technology can easily analyze code to obtain the functions, lines, and even the paths executed by code in test suites. For certain classes of programming errors such as typographical errors, the incremental code coverage is an ideal covariate for estimating the probability of detecting an error. The execution frequency of blocks of code or functions is clearly related to the probability of error detection. Figure 5 shows clearly that small portions of the program are heavily exercised but that most of the code is not touched. In an indirect way operational profile testing attempts to capture this idea by testing the features, and therefore the code, in relation to how often they will be used. This notion suggests that statistical techniques involving covariates can improve the efficiency of software testing.

Figure 7 suggests novel ways of displaying software metrics. The current practice is to identify overly complex files for special care and management attention. The procedures for identifying complex code are often based on very clever and sophisticated arguments, but not on data. A statistical approach might attempt to correlate the complexity of code with the locations of past faults and investigate their predictive power. Statistical models that can relate complexity metrics to actual faults will increase the models' practical efficiency for real-life systems. These models should not be developed in the absence of data about the code. Simple ways of presenting such data, such as an ordered list of fault density, file by file, can be very effective in guiding the selection of an appropriate model. In other cases, microanalysis, often driven by graphical browsers, might suggest a richer class of models that the data could support. For example, software fault rates are often quoted in terms of the number of faults per 1,000 lines of NCSL. The lines in Figure 1 can be color-coded to show the historical locations of past faults. In other representations (not shown), clear spatial patterns with faults are concentrated in particular files and in particular regions of the files, suggesting that spatial models of fault density might work very well in helping to identify fault-prone code.

Challenges for Visualization

The research opportunities and challenges in visualizing software data are similar to those for visualizing other large abstract databases:

1. Software data are abstract; there is no natural two-dimensional or three-dimensional representation of the data. A research challenge is to discover meaningful representations of the data that enable an analyst to understand the data in context.
2. Much software data are nontraditional statistical data such as the change history of source code, duplication in manuals, or the structure of a relational database. New metaphors must be discovered for harmonious transfer information.
3. The database associated with large software systems may be huge, potentially containing millions of observations. Effective statistical graphics techniques must be able to cope with the volume of data found in modern software systems.
4. The lack of easy-to-use software tools makes the development of high-quality custom visualizations particularly difficult. Currently, visualizations must be hand-coded in low-level languages such as C or C++. This is a time-consuming task that can be carried out only by the most sophisticated programmers.

Opportunities for Visualization

Visualizations associated with software involve the code itself, data associated with the system, the execution of the program, and the process for creating the system. Opportunities include the following:

1. **Objects/Patterns.** Object-oriented programming is rapidly becoming standard for development of new systems and is being retrofitted into existing systems. Effective

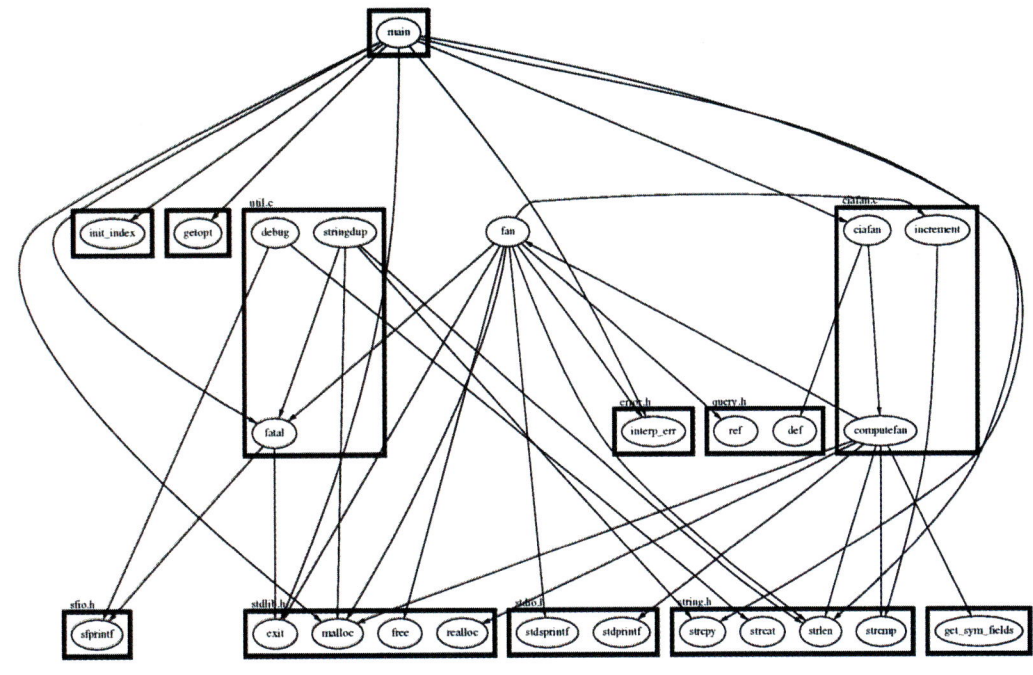
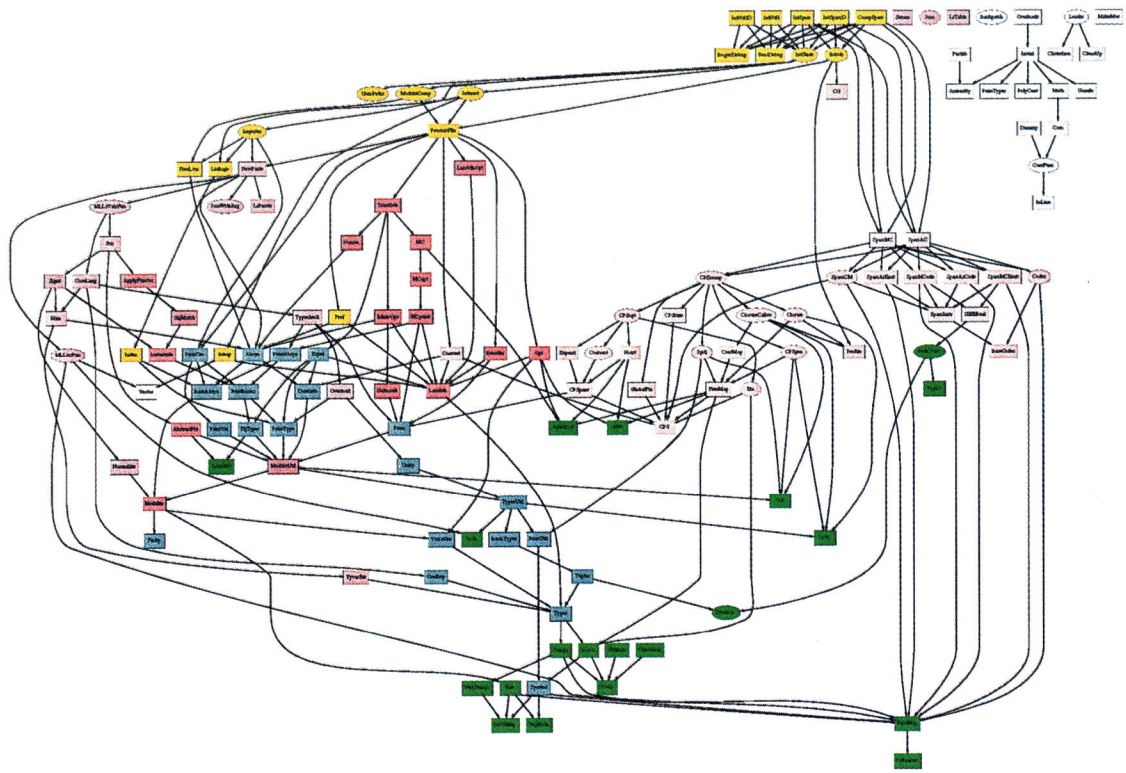

Figure 5. Function call graphs showing the calling pattern between procedures. The top panel shows an interpretable, easy-to-comprehend display, whereas the bottom panel is overly busy and visually confusing.

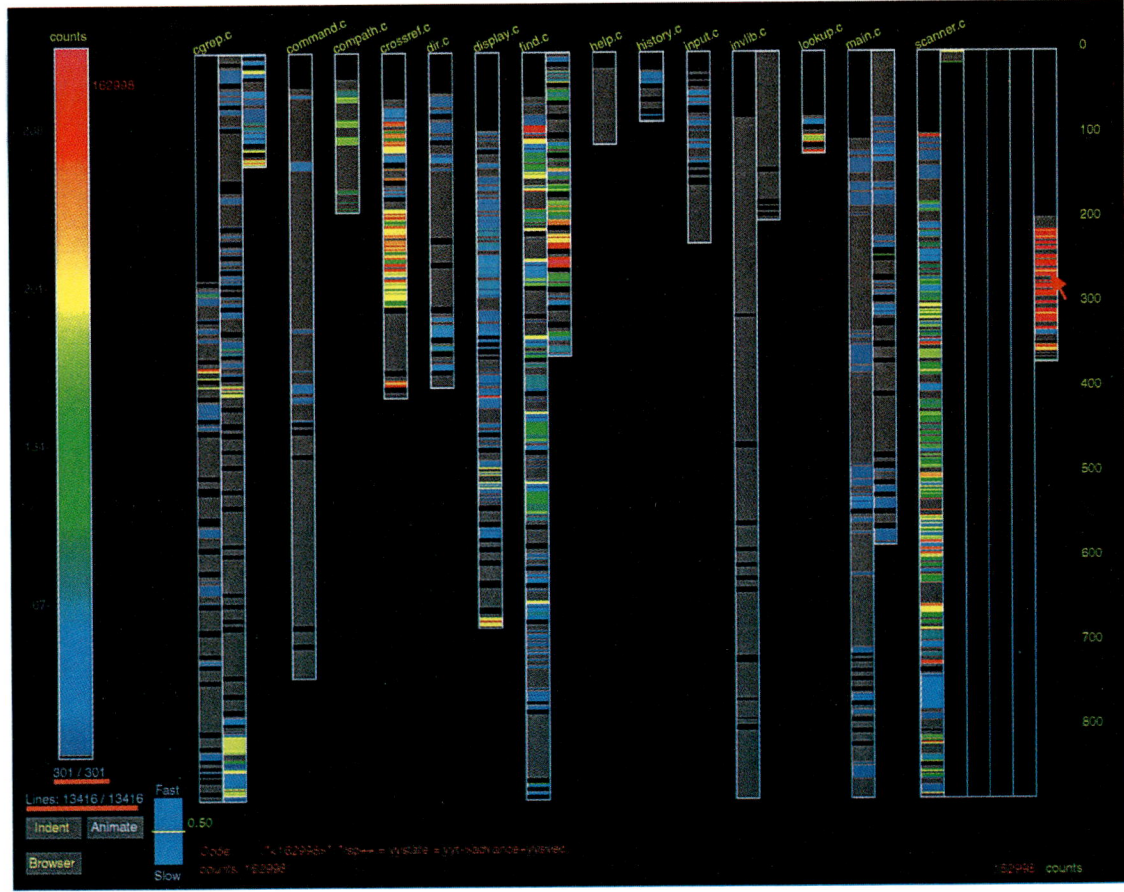

Figure 6. A *SeeSoft*™ display showing code coverage for a program executing its regression test. The color of each line is determined by the number of times that it executed. The colors range from red (the "hot spots") to deep blue (for code executed only once) using a red-green-blue color spectrum. There are two special colors: the black lines are non-executable lines of code such as variable declarations and comments, and the gray lines are the non-executed (not covered) lines. The figure shows that generating regression tests with high coverage is quite difficult. Source: Eick (1994).

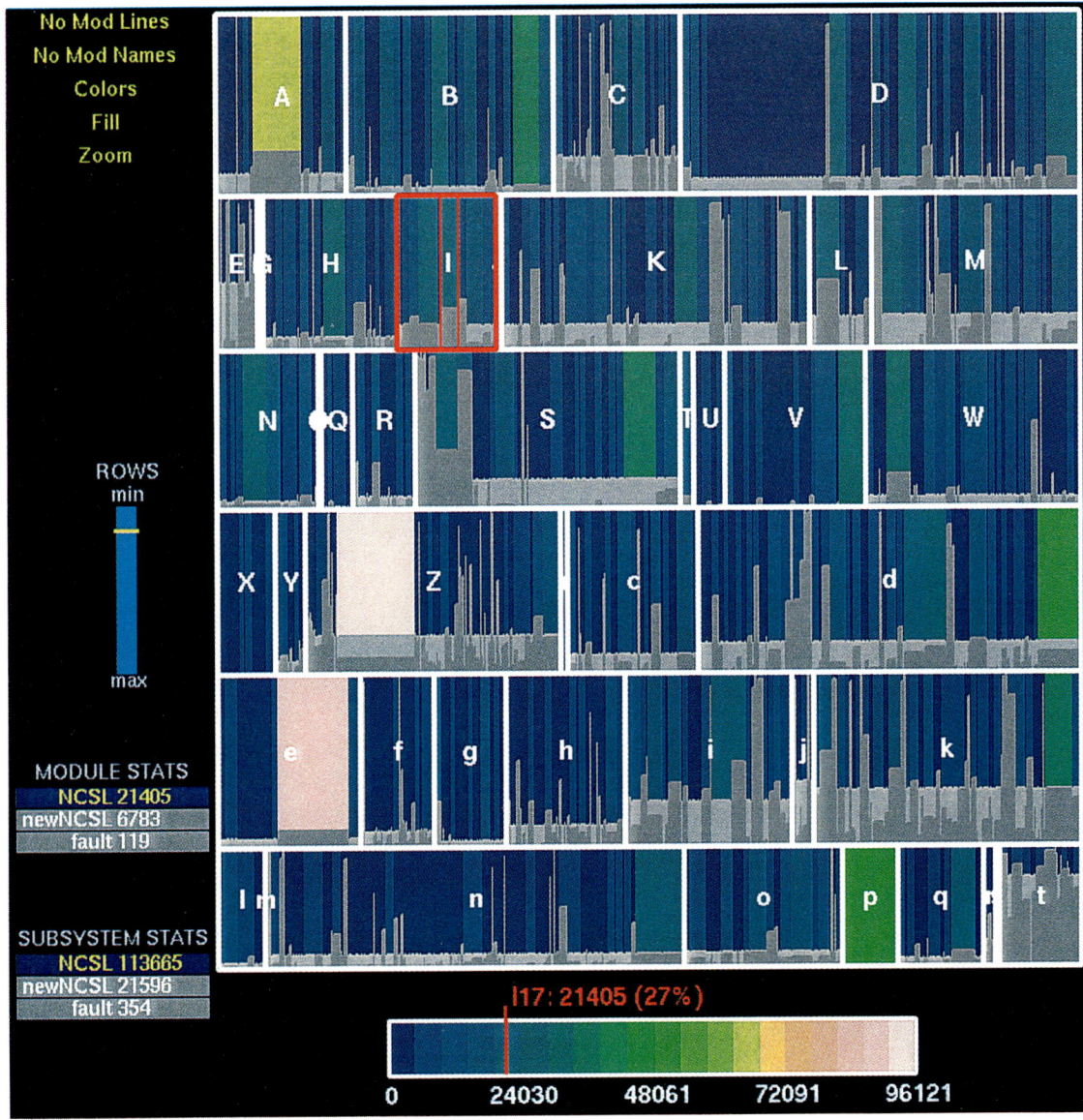

Figure 7. A display of software metrics for a million-line system. The rectangle forming the outermost boundary represents the entire system. The rectangles contained within the boundary represent the size (in NCSLs) of individual subsystems (each labeled with a single character A-Z, a-t), and modules within the subsystems. Color is used here to redundantly encode size according to the color scheme in the slider at the bottom of the screen.

displays need to be developed for understanding the inheritance (or dependency) structure, semantic relationships among objects, and the run-time life cycle of objects.

2. **Performance.** Software systems inevitably run too slowly, making run-time performance an important consideration. Host systems often collect large volumes of fine-grain (that is, low-level) performance data including function calling patterns, line execution counts, operating system page faults, heap usage, and stack space, as well as disk usage. Novel techniques to understand and digest dynamic program execution data would be immediately useful.

3. **Parallelism.** Recently, massively parallel computers with tens to thousands of cooperating processors have started to become widely available. Programming these computers involves developing new distributed algorithms that divide important computations among the processors. Most often an essential aspect of the computation involves communicating interim results between processors and synchronizing the computations. Visualization techniques are a crucial tool for enabling programmers to model and debug subtle computations.

4. **Three-dimensional.** Workstations capable of rendering realistic three-dimensional displays are rapidly becoming widely available at reasonable prices. New visualization techniques leveraging three-dimensional capabilities should be developed to enable software engineers to cope with the ever-increasing complexity of modern software systems.

ORTHOGONAL DEFECT CLASSIFICATION

The primary focus of software engineering is to monitor a software development process with a view toward improving quality and productivity. For improving quality, there have been two distinct approaches. The first considers each defect as unique and tries to identify a cause. The second considers a defect as a sample from an ensemble to which a formal statistical reliability model is fitted. Chillarege et al. (1992) proposed a new methodology that strikes a balance between these two ends of spectrum. This method, called orthogonal defect classification, is based on exploratory data analysis techniques and has been found to be quite useful at IBM. It recognizes that the key to improving a process is to quantify various cause-and-effect relationships involving defects.

The basic approach is as follows. First, classify defects into various types. Then, obtain a distribution of the types across different development phases. Finally, having created these reference distributions and the relationships among them, compare them with the distributions observed in a new product or release. If there are discrepancies, take corrective action.

Operationally, the defects are classified according to eight "orthogonal" (mutually exclusive) defect types: functional, assignment, interface, checking, timing, build/package/merge, data structures and algorithms, and documentation. Further, development phases are divided into four basic stages (where defects can be observed): design, unit test, function test, and system test. For each stage and each defect type, a range of acceptable baseline defect rates is defined by experience. This information is used to improve the quality of a new product or release. Toward this end, for a given defect type, defect distributions across development stages are compared with the baseline rates. For each chain of results—say, too high early on, lower later, and high at the end—an implication is derived. For example, the implication may be that function testing should be revamped.

This methodology has been extended to a study of the distribution of triggers, that is, the conditions that allow a defect to surface. First, it is implicit in this approach that there is no substitute for a good data analysis. Second, assumptions clearly are being made about the stationarity of reference distributions, an approach that may be appropriate for a stable environment with similar projects. Thus, it may be necessary to create classes of reference distributions and classes of similar projects. Perhaps some clustering techniques may be valuable in this context. Third, although the defect types are mutually exclusive, it is possible that a fault may result in many defects, and vice versa. This multiple-spawning may cause serious implementation difficulties. Proper measurement protocols may diminish such multipropagation. Finally, given good-quality data, it may be possible to extend orthogonal defect classification to efforts to identify risks in the production of software, perhaps using data to provide early indicators of product quality and potential problems concerning scheduling. The potential of this line of inquiry should be carefully investigated, since it could open up an exciting new area in software engineering.

6
Summary and Conclusions

In the 1950s, as the production line was becoming the standard for hardware manufacturing, Deming showed that statistical process control techniques, invented originally by Shewhart, were essential to controlling and improving the production process. Deming's crusade has had a lasting impact in Japan and has changed its worldwide competitive position. It has also had a global impact on the use of statistical methods, the training of statisticians, and so forth.

In the 1990s the emphasis is on software, as complex hardware-based functionality is being replaced by more flexible, software-based functionality. Small programs created by a few programmers are being superseded by massive software systems containing millions of lines of code created by many programmers with different backgrounds, training, and skills. This is the world of so-called software factories. These factories at present do not fit the traditional model of (hardware) factories and more closely resemble the development effort that goes into designing new products. However, with the spread of software reuse, the increasing availability of tools for automatically capturing requirements, generating code and test cases, and providing user documentation, and the growing reliance on standardized tuning and installation processes and standardized procedures for analysis, the model is moving closer to that of a traditional factory. The economy of scale that is achievable by considering software development as a manufacturing process, a factory, rather than a handcrafting process, is essential for preserving U.S. competitive leadership. **The challenge is to build these huge systems in a cost-effective manner**. The panel expects this challenge to concern the field of software engineering for the rest of the decade. Hence, **any set of methodologies that can help in meeting this challenge will be invaluable. More importantly, the use of such methodologies will likely determine the competitive positions of organizations and nations involved in software production.**

With the amount of variability involved in the software production process and its many subprocesses, as well as the diversity of developers, users, and uses, it is unlikely that a deterministic control system will help improve the software production process. As in statistical physics, only a technology based on statistical modeling, something akin to statistical control, will work. The panel believes that the juncture at hand is not very different from the one reached by Deming in the 1950s when he began to popularize the concept of statistical process control. **What is needed now is a detailed understanding by statisticians of the software engineering process, as well as an appreciation by software engineers of what statisticians can and cannot do**. If collaborative interactions and the building of this mutual understanding can be cultivated, then there likely will occur a major impact of the same order of magnitude as Deming's introduction of statistical process control techniques in hardware manufacturing.

Of course, this is not to say that all software problems are going to be solved by statistical means, just as not all automobile manufacturing problems can be solved by statistical means. On the contrary, the software industry has been technology driven, and the bulk of future gains in productivity will come from new, creative ideas. For example, much of the gain in productivity between 1950 and 1970 occurred because of the replacement of assembler coding by high-level languages.

Nevertheless, as the panel attempts to point out in this report, increased collaboration between software engineers and statisticians holds much promise for resolving problems in software development. Some of the catalysts that are essential for this interaction to be productive, as well as some of the related research opportunities for software engineers and statisticians, are discussed below.

INSTITUTIONAL MODEL FOR RESEARCH

The panel strongly believes that **the right model for statistical research in software development is collaborative in nature**. It is essential to avoid solving the "wrong" problems. It is equally important that the problems identified in this report not be "solved" by statisticians in isolation. Statisticians need to attain a degree of credibility in software engineering, and such credibility will not be achieved by developing N new reliability models with high-power asymptotics. **The ideal collaboration partners statisticians and software engineers in work aimed at improving a *real* software process or product**.

This conclusion assumes not only that statisticians and software engineers have a mutual desire to work together to solve software engineering problems, but also that funding and reward mechanisms are in place to stimulate the technical collaboration. Up to now, such incentives have not been the norm in academic institutions, given that, for example, coauthored papers have been generally discounted by promotion evaluation committees. Moreover, at funding agencies, proposals for collaborative work have tended to "fall through the cracks" because of a lack of interdisciplinary expertise to evaluate their merits. The panel expects such barriers to be reduced in the coming years, but in the interim, **industry can play a leadership role in nurturing collaborations between software engineers and statisticians** and can reduce its own set of barriers (for instance, those related to proprietary and intellectual property interests).

MODEL FOR DATA COLLECTION AND ANALYSIS

As discussed above in this report, for statistical approaches to be useful, it is essential that high-quality data be available. Quality includes measuring the right things at the right time—specifically, adopted software metrics must be relevant for each of the important stages of the development life cycle, and the protocol of metrics for collecting data must be well defined and well executed. **Without careful preparation that takes account of all of these data issues, it is unlikely that statistical methods will have any impact on a given software project under study. For this reason, it is crucial to have the software industry take a lead position in research on statistical software engineering**.

Figure 8, a model for the interaction between researchers and the software development process, displays a high-level spiral view of the software development process offered by Dalal

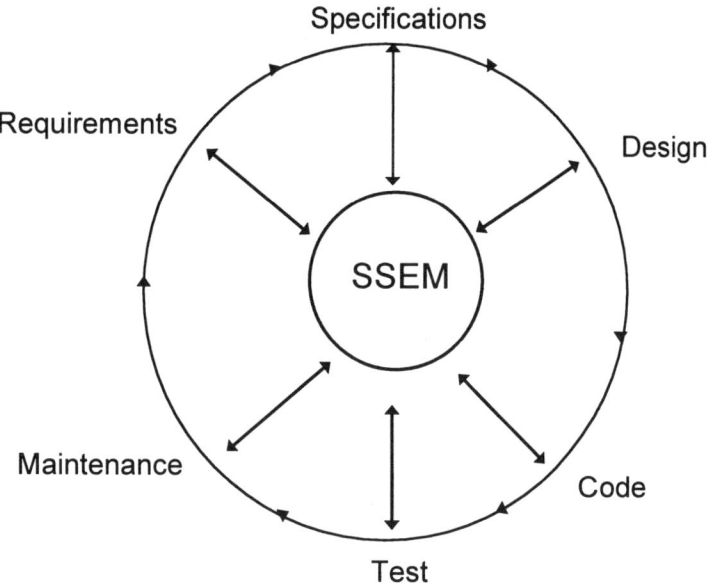

Figure 8. Spiral software development process model. SSEM, statistical software engineering module.

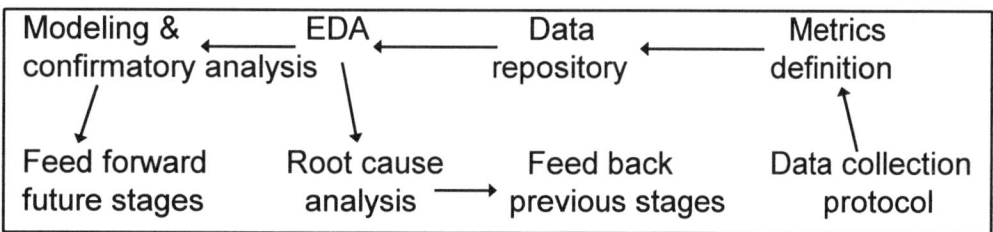

Figure 9. Statistical software engineering module at stage n.

et al. (1994). Figure 9 gives a more detailed view of the statistical software engineering module (SSEM) at the center of Figure 8.

The SSEM has several components. One of its major functions is to act as the central repository for all relevant project data (statistical or nonstatistical). Thus this module serves as a resource for the entire project, interfacing with every stage, typically at its review or conclusion. For example, the SSEM would be used at the requirement review stage, when data on inspection, faults, times, effort, and coverage are available. For testing, information would be gathered at the end of each stage of testing (unit, integration, system, alpha, beta, . . .) about the number of open faults, closed faults, types of problems, severity, changes, and effort. Such data would come from test case management systems, change management systems, and configuration management systems.

Additional elements of the SSEM include collection protocols, metrics, exploratory data analysis (EDA), modeling, confirmatory analysis, and conclusions. A critical part of the SSEM would be related to root-cause analysis. Analysis could be as simple as Ishikawa's fish bone diagram (Ishikawa, 1976), or more complex, such as orthogonal defect classification (described in Chapter 5). This capability accords with the belief that a careful analysis of root cause is essential to improving the software development process. Central placement of the SSEM ensures that the results of various analyses will be communicated at all relevant stages. For example, at the code review stage, the SSEM can suggest ways of improving the requirement process as well as point out potentially error-prone parts of the software for testing.

ISSUES IN EDUCATION

Enormous opportunities and many potential benefits are possible if the software engineering community learns about relevant statistical methods and if statisticians contribute to and cooperate in the education of future software engineers. The areas outlined below are those that are relevant today. As the community matures in its statistical sophistication, the areas themselves should evolve to reflect the maturation process.

- **Designed experiments.** Software engineering is inherently experimental, yet relatively few designed experiments have been conducted. Software engineering education programs must stress the desirability, wherever feasible, of validating new techniques through the use of statistically valid, designed experiments. Part of the reason for the lack of experimentation in software engineering may involve the large variability in human programming capabilities. As pointed out in Chapter 5, the most talented programmer may be 20 times more productive than the least talented. This disparity makes it difficult to conduct experiments because the between-subject variability tends to overwhelm the treatment effects. Experimental designs that address broad variability in subjects should be emphasized in the software engineering curriculum. A similar emphasis should be given to random- and fixed-effects models with hierarchical structure and to distinguishing within- and between-experiment variability.

    There is also a role for the statistics profession in the development of guidelines for experiments in software engineering akin to those mandated by the Food and Drug Administration for clinical trials. These guidelines will require reformulation in the software engineering context with the possible involvement of various industry and academic forums, including the Institute of Electrical and Electronics Engineers, the American Statistical Association, and the Software Engineering Institute.

- **Exploratory data analysis.** It is important to appreciate the strengths and the limitations of available data by challenging the data with a battery of numerical, tabular, and graphical methods. Exploratory data analysis methods (e.g., Tukey, 1977; Mosteller and Tukey, 1977) are essentially "model free," so that investigators can be surprised by unexpected behavior rather than have their thinking constrained by what is expected. One of the attitudes toward statistical analysis that is important to convey is that of

$$data = fit + residual.$$

The iterative nature of improving the model fit by removing structure from the residuals must be stressed in discussions of statistical modeling.

- **Modeling**. The models used by statisticians differ dramatically from those used by nonstatisticians. The differences stem from advances in the statistical community in the past decade that effectively relax assumptions of linearity for nearly all classical techniques. This relaxation is obtained by assuming only local linearity and using smoothing techniques (e.g., splines) to regularize the solutions (Hastie and Tibshirani, 1990). The result is quite flexible but interpretable models that are relatively unknown outside the statistics community. Arguably these more recent methods lack the well-studied inferential properties of classical techniques, but that drawback is expected to be remedied in coming years. Educational information exchanges should be conducted to stimulate more frequent and wider use of such comparatively recent techniques.

- **Risk analysis.** Software systems are often used in conjunction with other software and hardware systems. For example, in telecommunications, an originating call is connected by switching software; however, the actual connection is made by physical cables, transmission cells, and other components. The mega systems thus created run our nation's telephone systems, stock markets, and nuclear power plants. Failures can be very expensive, if not catastrophic. Thus, it is essential to have software and hardware systems built in such a way that they can tolerate faults and provide minimal functionality, while precluding a catastrophic failure. This type of system robustness is related to so-called fault-tolerant design of software (Leveson, 1986).

    Risk analysis has played a key role in identifying fault-prone components of hardware systems and has helped in managing the risks associated with very complex hardware-software systems. A paradigm suggested by Dalal et al. (1989) for risk management for the space shuttle program and corresponding statistical methods are important in this context. For software systems, risk analysis typically begins with identifying programming styles, characteristics of the modules responsible for most software faults, and so on. Statistical analysis of root-cause data leads to a risk profile for a system and can be useful in risk reduction. Risk management also involves consideration of the probability of occurrence of various failure scenarios. Such probabilities are obtained either by using the Delphi method (e.g., Dalkey, 1972; Pill, 1971) or by analyzing historical data. One of the key requirements in failure-scenario analysis is to dynamically update information about the scenarios as new data on system behavior become available, such as a changing user profile.

- **Attitude toward assumptions**. As software engineers are aware, a major difference between statistics and mathematics is that for the latter, it matters only that assumptions be correctly stated, whereas for the former, it is essential that the prevailing assumptions be supported by the data. This distinction is important, but unfortunately it is often taken too literally by many who use statistical techniques. Tukey has long argued that what is important is not so much that assumptions are violated but rather that their effect on conclusions is well understood. Thus for a linear model, where the standard assumptions include normality, homoscedasticity, and independence, their importance to statements of inference is exactly in the opposite order. Statistics textbooks, courses, and consulting activities should convey the statistician's level of understanding of and perspective on the importance of assumptions for statistical inference methods.

- **Visualization**. The importance of plotting data in all aspects of statistical work cannot be overemphasized. Graphics is important in exploratory stages to ascertain how complex a model the data can support; in the analysis stage for display of residuals to examine what a currently entertained model has failed to account for; and in the presentation stage where graphics can provide succinct and convincing summaries of the statistical analysis and associated uncertainty. Visualization can also help software engineers cope with, and understand, the huge quantities of data collected in the software development process.

- **Tools**. Software engineers tend to think of statisticians as people who know how to run a regression software package. Although statisticians prefer to think of themselves more as problem solvers, it is still important that they point out good statistical computing tools—for instance, S, SAS, GLIM, RS1, and so on—to software engineers. A CATS report (NRC, 1991) attempts to provide an overview of statistical computing languages, systems, and packages, but for such material to be useful to software engineers, a more focused overview will be required.

References

Abdel-Ghaly, A.A., P.Y. Chan, and B. Littlewood. 1986. Evaluation of competing software reliability predictions. *IEEE Trans. Software Eng.* **SE-12**(9):950-967.

Abdel-Hamid, T. 1991. *Software Project Dynamics: An Integrated Approach.* Englewood Cliffs, N.J.: Prentice-Hall.

American Heritage Dictionary of the English Language, The. 1981. Boston: Houghton Mifflin.

American Statistical Association (ASA). 1993. *Combining Information: Statistical Issues and Opportunities for Research*, Contemporary Statistics Series, No. 1. Alexandria, Va.: American Statistical Association.

Baecker, R.M. and A. Marcus. 1988. *Human Factors and Typography for More Readable Programs.* Reading, Mass.: Addison Wesley.

Baker, M.J. and S.G. Eick. 1995. Space-filling software displays. *J. Visual Languages Comput.* **6**(2). In press.

Basili, V. 1993. Measurement, analysis and modeling, and experimentation in software engineering. Unpublished paper presented at Forum on Statistical Methods in Software Engineering, October 11-12, 1993, National Research Council, Washington, D.C.

Basili, V. and D. Weiss. 1984. A methodology for collecting valid software engineering data. *IEEE Trans. Software Eng.* **SE-10**:6.

Becker, R.A. and W.S. Cleveland. 1987. Brushing scatterplots. *Technometrics* **29**:127-142.

Becker, R.A., W.S. Cleveland, and A.R. Wilks. 1987. Dynamic graphics for data analysis. *Statistical Science* **2**:355-383.

Beckman, R.J. and M.D. McKay. 1987. Monte Carlo estimation under different distributions using the same simulation. *Technometrics* **29**:153-160.

Blum, M., M. Luby, and R. Rubinfeld. 1989. Program result checking against adaptive programs and in cryptographic settings. Pp. 107-118 in *Distributed Computing and Cryptography*, J. Feigenbaum and M. Merritt, eds. DIMACS: Series in Discrete Mathematics and Theoretical Computer Science, Vol. 2. Providence, R.I.: American Mathematical Society.

Blum, M., M. Luby, and R. Rubinfeld. 1990. Self-testing/correcting with applications to numerical problems. *STOC* **22**:73-83.

Boehm, B.W. 1981. *Software Engineering Economics.* Engelwood Cliffs, N.J.: Prentice Hall.

Brocklehurst, S. and B. Littlewood. 1992. New ways to get accurate reliability measures. *IEEE Software* **9**(4):34-42.

Brown, M.H. and J. Hershberger. 1992. Color and sound in algorithm animation. *IEEE Computer* **25**(12):52-63.

Burnham, K.P. and W.S. Overton. 1978. Estimation of the size of a closed population when capture probabilities vary among animals. *Biometrika* **45**:343-359.

Chillarege, R., I. Bhandari, J. Chaar, M. Halliday, D. Moebus, B. Ray, and M. Wong. 1992. Orthogonal defect classification—A concept for in-process measurements. *IEEE Trans. Software. Eng.* **SE-18**:943-955.

Cohen, D.M., S.R. Dalal, A. Kaija, and G. Patton. 1994. The automatic efficient test generator (AETG) system. Pp. 303-309 in *Proceedings of the 5th International Symposium on Software Reliability Engineering.* Los Alamitos, Calif.: IEEE Computer Society Press.

Curtis, W. 1988. The impact of individual differences in programmers. Pp. 279-294 in *Working with Computers: Theory versus Outcome,* G.C. van der Veer et al., eds. San Diego, Calif.: Academic Press.

Dalal, S.R. and C.L. Mallows. 1988. When should one stop software testing? *J. Am. Statist. Assoc.* **83**:872-879.

Dalal, S.R. and C.L. Mallows. 1990. Some graphical aids for deciding when to stop testing software. *IEEE J. Selected Areas in Communications* **8**:169-175. (Special issue on software quality and productivity.)

Dalal, S.R. and C.L. Mallows. 1992. Buying with exact confidence. *Ann. Appl. Probab.* **2**:752-765.

Dalal, S.R. and A.M. McIntosh. 1994. When to stop testing for large software systems with changing code. *IEEE Trans. Software Eng.* **SE-20**:318-323.

Dalal, S.R., E.B. Fowlkes, and A.B. Hoadley. 1989. Risk analysis of the space shuttle: Pre-Challenger prediction of failure. *J. Am. Stat. Assoc.* **84**:945-957.

Dalal, S.R., J.R. Horgan, and J.R. Kettenring. 1994. Reliable software and communication II: Controlling the software development process. *IEEE J. Selected Areas in Communications* **12**:33-39.

Dalkey, N.C. 1972. *Studies in the Quality of Life—Delphi and Decision-Making.* Lexington, Mass.: D.C. Heath & Co.

Dawid, A.P. 1984. Statistical theory: The prequential approach. *J. R. Stat. Soc. London A* **147**:278-292.

DeMillo, R.A., D.S. Guindi, K.S. King, W.M. McCracken, and A.J. Offutt. 1988. An extended overview of the MOTHRA mutation system. Pp. 142-151 in *Proceedings of the Second Workshop on Software Testing, Verification and Analysis.* Alberta, Canada: Banff.

Ebert, C. 1992. Visualization techniques for analyzing and evaluating software measures. *IEEE Trans. Software Eng.* **11**(18):1029-1034.

Eckhardt, D.E. and L.D. Lee. 1985. A theoretical basis of multiversion software subject to coincident errors. *IEEE Trans. Software Eng.* **SE-11**:1511-1517.

Eckhardt, D.E., A.K. Caglayan, J.C. Knight, L.D. Lee, D.F. McAllister, M.A. Vouk, and J.P. Kelly. 1991. An experimental evaluation of software redundancy as a strategy for improving reliability. *IEEE Trans. Software Eng.* **SE-17**(7):692-702.

Eick, S.G. 1994. Graphically displaying text. *J. Comput. Graphical Stat.* **3**(2):127-142.

Eick, S.G., C.R. Loader, M.D. Long, S.A. Vander Wiel, and L.G. Votta. 1992a. Estimating software fault content before coding. Pp. 59-65 in *Proceedings of the 14th International Conference on Software Engineering* (Melbourne, Australia). Los Alamitos, Calif.: IEEE Computer Society Press.

Eick, S.G., J.L. Steffen, and E.E. Sumner. 1992b. SeeSoft™—A tool for visualizing line oriented software. *IEEE Trans. Software Eng.* **11**(18):957-968.

Ganser, E.R., E.E. Koutsofios, S.C. North, and K.-P. Vo. 1993. A technique for drawing directed graphs. *IEEE Trans. Software Eng.* **SE-19**(3):214-230.

Halstead, M.H. 1977. *Elements of Software Science.* New York: Elsevier.

Hastie, T.J. and R.J. Tibshirani. 1990. *Generalized Additive Models.* London: Chapman & Hall.

Henrion, M. and B. Fischhoff. 1986. Assessing uncertainty in physical constants. *Am. J. Phys.* **54**(9):791-798.

Horgan, J.R. and S. London. 1992. ATAC: A data flow testing tool for C. Pp. 2-10 in *Proceedings of the Second Symposium on Assessment of Quality Software Development Tools* (May 27-29, 1992, New Orleans, La.), E. Nahouraii, ed. Los Alamitos, Calif.: IEEE Computer Society Press.

Humphrey, W.S. 1988. Characterizing the software process: A maturity framework. *IEEE Software* **5**:73-79.

Humphrey, W.S. 1989. *Managing the Software Process.* Reading, Mass.: Addison Wesley.

Iman, R.L. and W.J. Conover. 1982. A distribution free approach to inducing rank correlations among input variables. *Commun. Stat., Part B* **11**:311-334.

Institute of Electrical and Electronics Engineers (IEEE). 1990. *IEEE Standard Glossary of Software Engineering Terminology.* IEEE Std. 610.12-1990. New York: IEEE, Inc.

Institute of Electrical and Electronics Engineers (IEEE). 1993. *IEEE Standard for Software Productivity Metrics.* IEEE Computer Society, IEEE Std. 1045-1992, January 11, 1993. New York: IEEE, Inc.

Ishikawa, K. 1976. *Guide to Quality Control.* Tokyo, Japan: Asian Productivity Organization.

Kahneman, D., P. Slovic, and A. Tversky, eds. 1982. *Judgment Under Uncertainty: Heuristics and Biases.* New York: Cambridge University Press.

Keller, T.W. 1993. Maintenance process metrics for space shuttle flight software. Unpublished paper presented at Forum on Statistical Methods in Software Engineering, October 11-12, 1993, National Research Council, Washington, D.C.

Kitchenham, B. 1991. Never mind the metrics; what about the numbers! Pp. 28-37 in *Formal Aspects of Measurement,* T. Denvir, R. Herman, and R.W. Whitty, eds. Proceedings of the BCS-FACS Workshop, May 5, 1991, South Bank University, London. New York: Springer-Verlag.

Kitchenham, B. 1992. *Analyzing Software Data.* Metrics Club Report. Manchester, England: National Computing Centre, Ltd.

Knight, J.C. and N.G. Leveson. 1986. Experimental evaluation of the assumption of independence in multiversion software. *IEEE Trans. Software Eng.* **SE-12**(1):96-109.

Lee, D. and M. Yanakakis. 1992. On-line minimization of transition systems. Pp. 264-274 in *Proceedings of the 24th Annual ACM Symposium on Theory of Computing.* New York: Association for Computing Machinery.

Leveson, N.G. 1986. Software safety: why, what, and how. *ACM Comput. Surveys* **8**:125-163.

Lipton, R. 1989. New directions in testing. Pp. 191-202 in *Distributed Computing and Cryptography,* J. Feigenbaum and M. Merritt, eds. DIMACS: Series in Discrete Mathematics and Theoretical Computer Science, Vol. 2. Providence, R.I.: American Mathematical Society.

Littlewood, B. 1979. Software reliability model for modular program structure. *IEEE Trans. Reliability* **R-28**(3):241-246.

Littlewood, B. and D.R. Miller. 1989. Conceptual modeling of coincident failures in multi-version software. *IEEE Trans. Software Eng.* **SE-15**(12):1596-1614.

Littlewood, B. and L. Strigini. 1993. Validation of ultra-high dependability for software-based systems. *Communications of the Association for Computing Machinery* **36**(11):69-80.

Mallows, C.L. 1973. Some comments on Cp. *Technometrics* **15**:661-667.

McCabe, T.J. 1976. A complexity measure. *IEEE Trans. Software Eng.* **SE-1**(3):312-327.

McKay, M.D., W.J. Conover, and R.J. Beckman. 1979. A comparison of three methods for selecting values of input variables in the analysis of output from a computer code. *Technometrics* **21**:239-245.

Mosteller, F. and J.W. Tukey. 1977. *Data Analysis and Regression: A Second Course in Statistics*. Reading, Mass.: Addison Wesley.

Munson, J.C. 1993. The relationship between software metrics and quality metrics. Unpublished paper presented at Forum on Statistical Methods in Software Engineering, October 11-12, 1993, National Research Council, Washington, D.C.

National Research Council (NRC). 1991. *The Future of Statistical Software*. Committee on Applied and Theoretical Statistics, Board on Mathematical Sciences. Washington, D.C.: National Academy Press.

National Research Council (NRC). 1992. *Combining Information: Statistical Issues and Opportunities for Research*. Committee on Applied and Theoretical Statistics, Board on Mathematical Sciences. Washington, D.C.: National Academy Press. (Reprinted in 1993 by the American Statistical Association as Volume 1 in the ASA Contemporary Statistics series.)

Nayak, T.K. 1988. Estimating population size by recapture sampling. *Biometrika* **75**:113-120.

Phadke, M.S. 1993. Robust design method for software engineering. Unpublished paper presented at Forum on Statistical Methods in Software Engineering, October 11-12, 1993, National Research Council, Washington, D.C.

Pill, J. 1971. The Delphi method: Substance, context, a critique and an annotated bibliography. *Socio-Economic Planning Science* **5**:57-71.

Randell, B. and P. Naur, eds. 1968. *Software Engineering Concepts and Techniques*. NATO Science Committee, Proceedings of the NATO Conferences, October 7-11, 1968, Garmisch, Germany. New York: Petrocelli/Charter.

Sackman, H. 1970. *Man-Computer Problem-Solving: Experimental Evaluation of Time-Sharing and Batch Processing*. New York: Auerbach.

Siegrist, K. 1988a. Reliability of systems with Markov transfers of control. *IEEE Trans. Software Eng.* **SE-14**(7):1049-1053.

Siegrist, K. 1988b. Reliability of systems with Markov transfers of control, II. *IEEE Trans. Software Eng.* **SE-14**(10):1478-1480.

Singpurwalla, N.D. 1991. Determining an optimal time interval for testing and debugging software. *IEEE Trans. Software Eng.* **17**(4):313-319.

Smith, A.F.M. and G.O. Roberts. 1993. Bayesian computation via the Gibbs sampler and related Markov chain Monte Carlo methods. *J. R. Stat. Soc. London B* **55**(1):3-23.

Stasko, J. 1993. Software visualization. Unpublished paper presented at Forum on Statistical Methods in Software Engineering, October 11-12, 1993, National Research Council, Washington, D.C.

Stein, M. 1987. Large sample properties of simulations using Latin hypercube sampling. *Technometrics* **29**:143-151.

Tukey, J.W. 1977. *Exploratory Data Analysis*. Reading, Mass.: Addison Wesley.

Tukey, J.W. 1991. Use of many covariates in clinical trials. *Int. Stat. Rev.* **59**(2):123-128.

Vander Wiel, S.A. and L.G. Votta. 1993. Assessing software designs using capture-recapture methods. *IEEE Trans. Software Eng.* **SE-19**(11):1045-1054.

Zuse, H. 1991. *Software Complexity: Measures and Methods*. Berlin: de Gruyter.

Zweben, S. 1993. Statistical methods in a study of software re-use principles. Unpublished paper presented at Forum on Statistical Methods in Software Engineering, October 11-12, 1993, National Research Council, Washington, D.C.

Appendix: Forum Program

MONDAY, OCTOBER 11, 1993

8:00 AM   Welcome and Introductions

8:05 AM   *Session on **Software Process***

            Session Chair: **Gloria J. Davis** (NASA-Ames Research Center)

            Invited Speakers: **Ted W. Keller** (IBM Corporation)
                               David Card (Computer Sciences Corporation)

9:45 AM   Break

10:15 AM   *Session on **Software Metrics***

            Session Chair: **Bill Curtis** (Carnegie Mellon University)

            Invited Speakers: **Victor R. Basili** (University of Maryland)
                               John C. Munson (University of Florida)

NOON   Break

1:00 PM   *Session on **Software Dependability and Testing***

            Session Chair: **Richard A. DeMillo** (Purdue University)

            Invited Speakers: **John C. Knight** (University of Virginia)
                               Richard Lipton (Princeton University)

2:25 PM   Break

3:15 PM   *Session on **Case Studies***

            Session Chair: **Daryl Pregibon** (AT&T Bell Laboratories, Murray Hill)

            Invited Speakers: **Tsuneo Yamaura** (Hitachi Computer Products-America, Inc.)
                               Stuart Zweben (Ohio State University)

5:00 PM   Adjourn

TUESDAY, OCTOBER 12, 1993

8:00 AM   *Session on **Nonstandard Methods***

           Session Chair: **Siddhartha R. Dalal** (Bellcore)

           Invited Speakers: **Madhav S. Phadke** (Phadke Associates)
                            Eric E. Sumner, Jr. (AT&T Bell Laboratories-Naperville)

9:45 AM   Break

10:15 AM   *Session on **Software Visualization***

           Session Chair: **Stephen G. Eick** (AT&T Bell Laboratories-Naperville)

           Invited Speakers: **William Hill** (Bellcore)
                            John Stasko (Georgia Institute of Technology)

NOON   Adjourn